Charles Dickens

A Christmas Carol

Text adaptation, notes and
activities by **Peter Foreman**

Editors: Rosalba Foreman, Richard Elliott
Design and art direction: Nadia Maestri
Computer graphics: Simona Corniola
Illustrations: Anna and Elena Balbusso

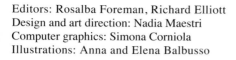

Picture Credits

By courtesy of the National Portrait Gallery, London: 4; bfi Stills: 8, 9;
Richard Green Gallery: 92

We would be happy to receive your comments and suggestions,
and give you any other information concerning our material.
http://publish.commercialpress.com.hk/blackcat/

CISQ CISQCERT
TEXTBOOKS AND
TEACHING MATERIALS
The quality of the publisher's
design, production and sales processes has
been certified to the standard of
UNI EN ISO 9001

ISBN 978962 07 0497 0

Printed in China by Leo Paper

Contents

Some information about Charles Dickens 5

Filmography 8

Chapter One | Scrooge 13

Chapter Two | Marley's Ghost 25

Chapter Three | The First Spirit 43

Chapter Four | The Second Spirit 62

Chapter Five | The Last of the Spirits 77

Chapter Six | 'A Merry Christmas, Mr Scrooge!' 96

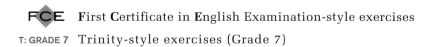

Dossiers | Some Christmas Ghosts 39

The Christmas Story 57

London in Dickens's Time 91

ACTIVITIES 20, 32, 52, 72, 86, 102

INTERNET PROJECT 94

EXIT TEST 107

FCE First Certificate in English Examination-style exercises

T: GRADE 7 Trinity-style exercises (Grade 7)

This story is recorded in full.

 These symbols indicate the beginning and end of the extracts linked to the listening activities.

Charles Dickens (1839) by Daniel Maclise.

Some information about Charles Dickens

Charles Dickens was born in Portsea in 1812, the son of John and Elizabeth Dickens. His father worked for the Navy Pay Office and for several years Charles enjoyed a happy childhood. These happy times did not last and when he was ten years old his family moved to London. His parents always had money problems and they sent Charles to work in a factory when he was twelve years old. He never forgot this humiliation and much of his later writing deals with themes such as the poor and social injustice. Later, he was separated from his parents when they were sent to prison because they couldn't pay back the money which they had borrowed.

At nineteen, he became a newspaper reporter, working for the *Mirror of Parliament* newspaper where he reported on electoral reforms such as the Reform Bill and the Factory Act. He married Kate Hogarth, the daughter of the chief editor of the *Evening Chronicle*, and they went on to have ten children.

He began to write sketches – short stories and descriptions of English life – for magazines. *Sketches by Boz*, published in 1836, was very popular. In the same year his first novel *The Pickwick Papers* appeared in a magazine every month, and it soon became a great success. Dickens finished it a year later, when Victoria became queen, and it was the first of many best-selling novels, published in parts every week or month. They include *Oliver Twist* (1837-8), *Nicholas Nickleby* (1854), *David Copperfield* (1849-50), *Hard Times* (1854), and *Great Expectations* (1860-1).

But Dickens did lots of other things. In his lifetime he was a magazine editor, an amateur actor, a writer of plays, articles, and hundreds of letters. He also liked travelling, and he lived in Italy (1844-5), Switzerland (1846-7), and in Paris (1847). In 1842 he toured North America, and he returned there in 1867-8 to read his books in public. This was an enormous success, but it was very hard work and he became ill. He died in 1870 and was buried in Westminster Abbey. [1]

---

1. **Westminster Abbey** : famous cathedral in central London.

Dickens's books were bestsellers and they are still popular. He always attacked the materialism of Victorian society and tried to show how it caused poverty and other social problems. *A Christmas Carol* is one example of Dickens's criticism of society and it is also the most famous Christmas story in the world.

1 **Decide whether the following statements are true (T) or false (F). Then correct the false ones.**

	T	F
a. Dickens wasn't born in London but he lived there.		
b. In 1824 Dickens's parents sent him to work in a factory.		
c. Later, Dickens went to prison with his parents.		
d. As a young man he worked for a newspaper.		
e. *Sketches by Boz* was his first novel.		
f. Dickens's novels were published in parts every week or month.		
g. Dickens had a very busy life and travelled a lot.		
h. The Americans did not like his books.		
i. People don't read Dickens's novels today.		
j. In *A Christmas Carol* Dickens criticised Victorian society.		

Filmography

A *Christmas Carol* has been filmed for the large and small screen over a hundred times. Here is a small selection from those titles.

A *Christmas Carol* directed by Edwin L. Marin, starring Reginald Owen, Gene Lockhart, Kathleen Lockhart (US, 1938)

Scrooge directed by Brian Desmond-Hurst, starring Alistair Sim, Michael Hordern, Kathleen Harrison (GB, 1951)

Scrooge directed by Ronald Neame, starring Albert Finney, Alec Guinness, Edith Evans (GB, 1970)

A scene from *Scrooge* (1951) featuring Glyn Dearman as Tiny Tim

A scene featuring Alistair Sim as Scrooge from *Scrooge* (1951)

Mickey's Christmas Carol directed by Burney Mattinson. Disney animated version (US, 1983)

A Christmas Carol directed by Clive Donner, starring George C. Scott, Frank Finlay, Angela Pleasance (GB, 1984)

Scrooged directed by Richard Donner, starring Bill Murray, Karen Allen, Robert Mitchum (US, 1988)

The Muppet Christmas Carol directed by Brian Henson, starring Michael Caine, Steven Mackintosh, Meridith Brown (US 1992)

A Christmas Carol directed by David Hugh Jones, starring Patrick Stewart, Richard E. Grant (US, 1999).

1 Read this newspaper article about *A Christmas Carol* and choose the answer (A, B, C, or D) which you think fits best according to the text.

The Morning Gazette

DECEMBER 30, 1843   PRICE 1d

From the Editor

A nation's thanks at Christmas

Mr Charles Dickens has written another bestseller and everybody is talking about it!

They say that when Mr Dickens wrote *A Christmas Carol*, he was very angry about the social condition of our nation's poor people. And he wanted to show us that the materialism and love of money in our society cause a lot of crime and poverty.

Well, Mr Dickens, the British people have understood your message. The book has been an enormous success.

When it was published a few days before Christmas, it sold 6,000 copies in twenty-four hours! A factory manager in America read it and he decided to give his workers an extra day's holiday!

Yes, *A Christmas Carol* has changed all of us, Mr Dickens! The whole country has tried to celebrate this Christmas with all the love, happiness and generosity that your wonderful story teaches.

The British nation thanks you!

1. Who wrote the article?
   A ☐ The British people.
   B ☐ A reader of *The Morning Gazette.*
   C ☐ The newspaper editor.
   D ☐ Charles Dickens.

2. According to the article, Dickens wanted to
   A ☐ write another bestseller.
   B ☐ show people that he was angry.
   C ☐ show people the causes of crime and poverty.
   D ☐ make a lot of money.

3. *A Christmas Carol* sold 6,000 copies

A ☐ in America.

B ☐ in twenty-four hours.

C ☐ before it was published.

D ☐ before Christmas.

4. How many extra days' holiday did the factory workers have?

A ☐ More than one.

B ☐ One.

C ☐ Seven.

D ☐ None.

5. The article thanks Dickens for

A ☐ showing people the true meaning of Christmas.

B ☐ his wonderful teaching.

C ☐ celebrating Christmas with love.

D ☐ changing the meaning of Christmas.

Before you read

1 **Match the words on the left to their definitions, like the example.**

a. the Christmas spirit   **1.** bright, colourful objects we put in the house

b. a Christmas card   **2.** a song about Christmas

c. Christmas Eve   **3.** kind, generous feelings

d. a Christmas carol   **4.** people say this at Christmas

e. Christmas decorations   **5.** a piece of illustrated card with a message inside that we send to people

f. 'a Merry Christmas!'   **6.** the day before Christmas Day

2 You are going to meet these characters in Chapter One. While you read it write their names below the pictures, and then say who they are (a-d).

1. ....

a. ....

2. ....

b. ....

3. ....

c. ....

4. ....

d. ....

Bob Cratchit	**Scrooge's nephew**
Scrooge	**a dead businessman**
Jacob Marley	**an office clerk**
Fred	**an old businessman**

Scrooge

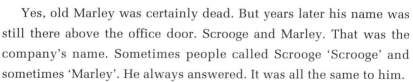

track 02

Marley was dead. That was certain because there were people at his funeral. Scrooge was there too. He and Marley were business partners, [1] and he was Marley's only friend. But Scrooge looked very happy at the funeral because on that day he made some money. Scrooge was a clever businessman.

Yes, old Marley was certainly dead. But years later his name was still there above the office door. Scrooge and Marley. That was the company's name. Sometimes people called Scrooge 'Scrooge' and sometimes 'Marley'. He always answered. It was all the same to him.

Oh, but he was a mean man, [2] Scrooge! He never spent any money and he never gave any away. He was an old miser. And he was a cold and solitary man. The cold was inside him. You could see it in his red eyes and on his blue nose and thin, white lips. You could hear it

---

1. **business** ['bɪznɪs] **partners** : together, Scrooge and Marley had a company for buying and selling things.
2. **mean man** : man who isn't generous.

in his hard voice, and it made his office cold, especially at Christmas. Nobody ever stopped him in the street to say, 'My dear Scrooge, how are you? When will you come and see me?' Children never spoke to him, and even dogs ran away from him. But Scrooge didn't care. [1] He liked it. That was what he wanted.

One Christmas Eve Scrooge was sitting in his office. It was only three o'clock in the afternoon but it was already dark. The weather was very cold and there was a lot of fog. It came into the office through the windows and doors. Bob Cratchit, Scrooge's clerk, [2] was copying letters in a dark little room, and the old man watched him carefully. Bob had a very very small fire in his room. It was even smaller than Scrooge's, and he tried to warm his hands at the candle but he couldn't do it.

'A merry [3] Christmas, uncle!' said a happy voice. And Scrooge's nephew [4] Fred came in.

'Bah!' answered Scrooge. 'Humbug!' [5]

His nephew looked warm. His face was red and his eyes were bright.

'Christmas a humbug, uncle?' he cried, surprised. 'You don't mean that, I'm sure.'

'Yes, I do,' said Scrooge. 'Merry Christmas! Why are you merry? You're a poor man, aren't you?'

'Well, why are *you* so unhappy? You're rich.'

'Bah! Humbug!'

---

1. **Scrooge didn't care** : it wasn't important to Scrooge.
2. **clerk** [kla:k] : person who works in an office or a bank.
3. **merry** : happy.
4. **nephew** ['nefju:] : (here) the son of Scrooge's sister.
5. **Humbug!** : (here) Scrooge is saying 'Don't talk nonsense!'

A Christmas Carol

'Don't be angry, uncle,' said Fred.

'Why not? There are too many fools [1] in this world. You say "Merry Christmas" when you're a year older and poorer. That's stupid!'

'Uncle – please!'

'Nephew! You have your own Christmas and I'll have mine. Leave me alone.'

'But you don't celebrate Christmas, uncle.'

'Because I never make any money at Christmas. I don't like it. Leave me alone.'

'But Christmas is a good time,' said the nephew. 'It's the only time in the year when people open their hearts and help each other. They become kind and generous. I like Christmas and I say God bless it!' [2]

The clerk in his little room clapped his hands [3] happily and said, 'Yes, that's right!'

'Another word from *you* and you'll lose your job,' Scrooge said to him.

'Don't be angry, uncle. Come and eat with us tomorrow,' said his nephew.

'No! Go away! I'm busy.' [4]

'But why won't you come?'

'Why did you get married?' Scrooge asked.

'Because I fell in love.'

'Because you fell in love! Bah! That's more stupid than a merry Christmas. Good afternoon.'

'But why don't you ever come to see me, uncle?'

---

1. **fools** : stupid people.
2. **God bless it!** : 'Thank God for Christmas!'
3. **clapped his hands** :
4. **busy** [bɪzi] : occupied.

'Good afternoon,' said Scrooge.

'Can't we be friends?'

'Good afternoon,' said Scrooge.

'Well, I'm very sorry about this, but I wish you [1] a merry Christmas with all my heart, uncle.'

'Good afternoon,' said Scrooge.

'And a happy new Year!'

'Good afternoon!' said Scrooge.

So his nephew went to the door and opened it. But before he left, he said 'Merry Christmas!' to the clerk, who answered with a warm 'Happy Christmas!'

'Are you stupid too?' Scrooge said.

At that moment two fat gentlemen came in.

'Excuse me, is this Scrooge and Marley's?' said one of them. 'May I ask if you are Mr Scrooge or Mr Marley?'

'Mr Marley is dead. He died on Christmas Eve seven years ago.'

'At this festive [2] time of the year, Mr Scrooge,' said the man, taking a pen from his pocket, 'we ask people to give some money to help the poor. There are thousands of people with nothing to eat at Christmas.'

'Aren't there any prisons?' asked Scrooge.

'Yes, lots of them.'

'And what about the workhouses? [3] Aren't there still lots of them?'

'Unfortunately, yes.'

'Good. I'm happy to hear it.'

'We don't think the people in the workhouses or prisons are

---

1. **I wish you** : I hope that you have.
2. **festive** : time of celebration.
3. **workhouses** : terrible government institutions for poor people without homes.

happy about it. They don't have much to eat or drink, and they're always cold. How much can you give us, sir?'

'Nothing!' Scrooge replied. 'Leave me alone. I don't celebrate Christmas and I don't give money to lazy people. [1] I help to pay for the workhouses and prisons. That's enough.'

'But many people can't go there and they'll die in this cold weather.'

'Well, there are too many people in the world already, so that's a good thing. Good afternoon, gentlemen!'

So the two men went out and Scrooge continued his work. It became colder and foggier [2] and darker. When a boy came to sing a Christmas carol [3] outside Scrooge's door, he stood up and shouted angrily, 'Go away!' The boy was frightened and ran away very quickly.

Finally, it was time to close the office and go home. Scrooge stopped his work and put down his pen. The clerk put on his hat to go.

'You want all day tomorrow, do you?' said Scrooge.

'If it's all right, sir – yes.'

'It's not all right,' Scrooge answered. 'I must pay you for a day's holiday.'

'It's only once a year, sir.'

'Bah! Every December 25th you get money for nothing! Well, arrive here extra early on the 26th – do you hear me?'

'Yes, sir,' said the clerk.

And when he left the office, he ran and danced all the way home because it was Christmas Eve.

1. **lazy people** : people who don't want to work.
2. **foggier** : (here) harder to see.
3. **carol** : Christmas song.

Understanding the text

1 **Are these sentences true (T) or false (F)? Correct the false ones, like the example.**

		T	F
a.	Scrooge didn't go to Marley's funeral.	☐	✓
b.	Scrooge and Marley worked together years ago.	☐	☐
c.	Scrooge always spent a lot of money.	☐	☐
d.	People liked Scrooge.	☐	☐
e.	Bob Cratchit and Fred were rich.	☐	☐
f.	Fred liked Christmas.	☐	☐
g.	Scrooge never visited his nephew.	☐	☐
h.	The two fat gentlemen were collecting money for the poor.	☐	☐
i.	Scrooge gave them some money.	☐	☐
j.	Marley died on December 25th.	☐	☐
k.	Scrooge didn't give any money to the carol singer.	☐	☐
l.	Bob Cratchit was happy because the next day was Christmas Day.	☐	☐

a. Scrooge went to Marley's funeral.
....
....
....
....
....
....
....
....
....
....
....

Past Continuous

One Christmas Eve Scrooge was sitting in his office.

The verb **was sitting** is in the Past Continuous (Past Simple of **to be** + present participle) and describes the continuation of an action in the past. Look at these examples.

*Early that morning the birds **were singing** in the garden.*
*Our cat **was watching** them.* (description)

*Scrooge **was working** in his office when his nephew Fred came in.* (the continuing action started before the simple past action and was interrupted by it.)

2 **Read the text and underline the appropriate words in italics.**

It was Christmas Eve. Scrooge **1** *was sitting/sat* in his office. The cold and fog **2** *came/was coming* in through the door and windows. A small fire **3** *burnt/was burning* in the fireplace. Bob Cratchit **4** *was copying/copied* letters in his dark room when suddenly his very small fire **5** *was going/went* out. He **6** *tried/was trying* to warm his cold hands at the candle, but he couldn't. Scrooge **7** *watched/was watching* him carefully when the door **8** *was opening/opened* and Scrooge's nephew Fred **9** *came/was coming* in.

3 Match the sentences in A and B to make one sentence with *when*.

A

a. ☐ On Christmas Eve Scrooge was working in his office.

b. ☐ A small fire was burning in the fireplace.

c. ☐ Bob Cratchit was trying to warm his hands at the candle.

d. ☐ Two fat gentlemen came in.

e. ☐ The boy was singing a Christmas carol.

when...

B

1. A voice said, 'A Merry Christmas!'

2. Scrooge was saying to Bob, 'Are you stupid too!'

3. His nephew Fred came in.

4. Scrooge shouted, 'Go away!'

5. It suddenly went out.

Listening

FCE **4** Listen to the beginning of this chapter and complete the sentences.

track 02

Marley was certainly dead because **1** ▭ at his funeral.

Years later Marley's name was still above **2** ▭

'Scrooge and Marley' was **3** ▭

You could hear Scrooge's coldness in his **4** ▭

People never spoke to Scrooge, but he **5** ▭
(2 possibilities)

On Christmas Eve it was already dark at only **6** ▭
in the afternoon.

The fog came into the office through the **7** ▭

Bob Cratchit worked as **8** ▭

Bob's fire was even **9** ▭

Bob tried to **10** ▭ at the candle.

Now read the text and check your answers.

5 **Listen again and tick (✓) the things that are mentioned.**

track 02

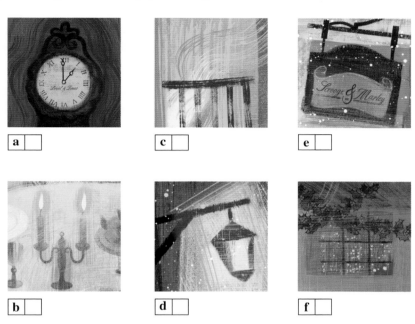

a ☐

c ☐

e ☐

b ☐

d ☐

f ☐

T: GRADE 7

6 Topic – Celebrations

Is Christmas celebrated in your country? Find some information or a picture showing what you do to celebrate. What aspect of Christmas does your picture/information show? Tell the class about it using these questions to help you.

a. Describe what the weather is usually like at Christmas where you live.

b. What do you usually eat for Christmas lunch?

c. Do you sometimes go on holiday for Christmas? Where?

d. What do you like about Christmas? What don't you like?

Before you read

 1 **For questions 1-12, read the text below and decide which answer (A, B, C or D) best fits each space. There is an example at the beginning (0).**

Scrooge walked home (**0**) *C* the rooms where he lived. Years ago (**1**) .....
partner Marley lived there. They were very old and dark and silent. The
knocker on the door was large, but it was like hundreds of other (**2**) .....
knockers. Scrooge (**3**) ..... looked at it. And he wasn't thinking about
Marley when he (**4**) ..... his key in the door. So how did he see Marley's
face in the knocker? Yes, Marley's face! There was a strange light
around (**5**) ..... . It looked at Scrooge with its glasses up in its hair, like
Marley (**6**) ..... he was alive. The hair (**7**) ..... moving slowly, the eyes
were wide open, and the face was very white. Scrooge looked at it for a
moment, and (**8**) ..... it was a knocker again. He was surprised, (**9**) .....
he went in and lit his candle. Then he looked at the knocker again.
'Pooh, pooh!' he said, and (**10**) ..... the door.
The sound echoed around the house, but Scrooge wasn't frightened
of echoes and he went slowly (**11**) ..... the dark stairs. He liked
darkness; it was cheap. He looked around his room: nobody under
the table, nobody under the sofa, nobody under the bed, nobody in
the cupboards. He locked the door and (**12**) ..... on his dressing-
gown, slippers and nightcap.

0.	**A** by	**B** at	**C** to	**D** in
1.	**A** her	**B** your	**C** their	**D** his
2.	**A** doors	**B** door's	**C** door	**D** doors'
3.	**A** didn't	**B** ever	**C** never	**D** hasn't
4.	**A** put	**B** took	**C** pulled	**D** locked
5.	**A** them	**B** us	**C** you	**D** it
6.	**A** where	**B** when	**C** what	**D** why
7.	**A** were	**B** are	**C** was	**D** been
8.	**A** then	**B** so	**C** as	**D** after
9.	**A** because	**B** after	**C** so	**D** but
10.	**A** unlocked	**B** closed	**C** opened	**D** knocked
11.	**A** along	**B** up	**C** in	**D** around
12.	**A** take	**B** wear	**C** put	**D** carry

Now read the beginning of Chapter Two and check your answers.

Marley's Ghost

track 03

Scrooge walked home to the rooms where he lived. Years ago his partner Marley lived there. They were very old and dark and silent. The knocker [1] on the door was large but it was like hundreds of other door knockers. Scrooge never looked at it. And he wasn't thinking about Marley when he put his key in the door. So how did he see Marley's face in the knocker? Yes, Marley's face! There was a strange light around it. It looked at Scrooge with its glasses up in its hair, like Marley when he was alive. The hair was moving slowly, the eyes were wide open, and the face was very white. Scrooge looked at it for a moment, and then it was a knocker again. He was surprised, but he went in and lit his candle. [2] Then he looked at the knocker again.

'Pooh, pooh!' [3] he said, and closed the door.

---

1. **knocker** : piece of metal on an outside door (of a house, an office etc.). You use it to attract the attention of the people inside.
2. **lit his candle** : made the candle give light (the past of the verb 'to light').
3. **Pooh, pooh!** : (here) Scrooge is saying 'It's nothing and I don't believe it.'

A Christmas Carol

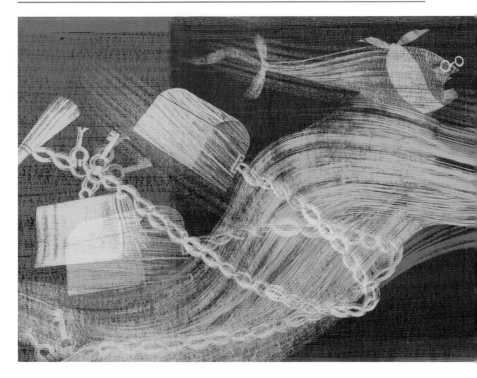

The sound echoed around the house, but Scrooge wasn't frightened of echoes and he went slowly up the dark stairs. He liked darkness; it was cheap. He looked around his room: nobody under the table, nobody under the sofa, nobody under the bed, nobody in the cupboards. 1 He locked the door and put on his dressing-gown, slippers and nightcap. 2 Then he sat in front of an old fireplace with a very small fire in it. For a moment he thought he saw Marley's face in the fire.

1. **cupboards** ['kʌbədz] : pieces of furniture (with doors) to keep clothes, plates or other things.
2. **dressing-gown, slippers ... nightcap** : long, warm piece of clothing, like a coat, to wear over pyjamas ... light, soft shoes to wear in the house ... long, soft hat that people wore in bed many years ago.

'Humbug!' he said.

Then he looked at the old bell [1] above him on the wall. He was very surprised when this bell began to move. At first it moved slowly and quietly, but soon it made a very loud sound and all the bells in the house began to ring too. Suddenly they stopped. Scrooge heard a strange noise far away in the house – a noise of metal, like chains. [2] It was coming up the stairs. Something was coming towards his door.

'It's humbug!' he said. 'I don't believe it.'

But the thing came into the room and stopped in front of him.

1. **bell** :   2. **chains** :

He couldn't believe his eyes! The same face: Marley's face! Scrooge recognised his dead partner's clothes and boots, and he saw a long chain round his transparent body. The chain had heavy cash-boxes, keys, locks, and account books [1] on it. Marley was looking at him with cold, dead eyes. There was a handkerchief [2] round his head and chin. [3]

'Well?' Scrooge said. 'What do you want with me?'

'Much!' It was certainly Marley's voice.

'Who are you?'

'Ask me who I *was*?'

'Who *were* you then?'

'In life I was your partner, Jacob Marley.'

'Sit down – if you can.'

The Ghost [4] sat in a chair on the other side of the fireplace.

'You don't believe in me, do you?' it said.

'No, I don't.'

'Why not?'

'Because perhaps I ate a piece of meat or cheese and my stomach didn't digest it, so you are only the consequence of a bad stomach.'

Scrooge said this because he didn't want to show his terror. But the Ghost's cold eyes frightened him very much.

'If I eat this candle,' Scrooge continued, 'I'll see hundreds of ghosts like you, but they'll only be in my head.'

---

1. **cash-boxes ... account books** : boxes containing money ... books for writing sums of money paid and received.
2. **handkerchief** ['hæŋkətʃif] : small piece of material to clean the nose.
3. **chin** :
4. **Ghost** : spirit of a dead person.

Then the Spirit gave a terrible cry, [1] and it shook [2] its chain with a tremendous noise. Scrooge trembled. And then he fell [3] out of his chair with horror when the Ghost took off the handkerchief and its chin dropped on its chest. [4]

'Help!' he cried with his hands on his face. 'Oh, why are you here, terrible Spirit?'

'Do you believe in me or not?'

'Yes, I do – I must!' Scrooge replied. 'But why do you come to me?'

'If a man's spirit stays away from other people while he is alive, it must walk through the world after he is dead, but it cannot share [5] the happiness of living people.' And again the Ghost shook its chain with a sad cry.

'Why are you wearing that chain?' Scrooge asked, trembling.

'Because I made it when I was alive. I stayed away from other people. I didn't try to help them. I never loved anybody; I loved only money. So I made this chain for myself and now I must wear it. I lived like you, Scrooge! Seven years ago your chain was long and heavy. Now it is very long and very heavy!'

Again Scrooge trembled in terror. 'Tell me more, old Jacob Marley. Help me!'

'I cannot help you, Ebenezer Scrooge,' answered the Ghost. 'I cannot rest, I cannot stay here. When I was alive, my spirit never

---

1. **cry** : (here) scream, loud noise.
2. **shook** [ʃʊk] : moved quickly up and down and from side to side (the past of the verb 'to shake').
3. **fell** : dropped to the floor (the past of the verb 'to fall').
4. **chest** : front part of the body between the shoulders.
5. **share** : (here) participate in.

walked out of our office. It was locked in there while I made all my money. So now I must travel and never stop.'

'Have you travelled all this time – for seven years?'

'Yes. No rest. No peace. Always travelling.'

'Do you travel fast?'

'Very fast. Like the wind.'

'Well, in seven years you have been to a lot of places then.'

'Oh, but I am a prisoner!' cried the phantom, [1] and it shook the chain again, a terrible sound in the silence of the night. 'I was also a prisoner in my life because I didn't try to help others.'

'But you were a good man of business, Jacob.' Scrooge was thinking of himself too.

'Business! What was my business? My business was people, my business was charity, [2] my business was love, my business was goodness! But I didn't do anything good. I lived with my eyes closed. I didn't see the poor and hungry people in the streets. But now I must go. Listen!'

'I'm listening, Jacob,' Scrooge said.

'I am here tonight to tell you something. There is still hope for you, Ebenezer. You still have a chance.'

'You were always a good friend, Jacob. Thank you.'

'You will see three Ghosts.'

Scrooge looked frightened. 'Are they the hope and the chance you spoke about, Jacob?'

'Yes.'

'Well – I don't want to see them...'

---

1. **phantom** : ghost.
2. **charity** : voluntary gift of help or money.

'You must! If you don't want to be like me, you must! The first Spirit will come at one o'clock tomorrow morning.'

'Can't they *all* come at one o'clock and finish it quickly, Jacob?'

'The second will come on the next night at the same time. The third will come on the night after that when the church bell strikes [1] twelve midnight. You will not see me again. Remember my words!'

Then the Ghost put the handkerchief round its head and began to walk towards the window. It asked Scrooge to follow. But when the window opened, Scrooge stopped. He was very frightened because he could hear a great noise of crying outside. The air was full of ghosts. They were moving quickly here and there, and they all wore chains like Marley's Ghost. Their cries were very sad. There was one old ghost with a big metal box of money on a chain. It was unhappy because it couldn't help a poor woman and her baby out in the cold, foggy night without a home.

Marley's Ghost went out into the night. In a moment it was with the other ghosts, and all of them disappeared. [2] Scrooge closed the window and went to the door. It was locked. [3] Did Marley's Ghost really come through a locked door?

'Bah!' he said. And he began to say 'Humbug!' but stopped. He didn't want to say it now.

It was late and he was tired. So he went to bed and fell asleep immediately.

---

1. **strikes** ['straɪks] : (here) rings to indicate the time.
2. **disappeared** [dɪsə'pɪəd] : vanished.
3. **locked** : closed with a key.

Understanding the text

1 **Answer these questions.**

   a. What was Scrooge doing when he saw Marley's face for the first time?

   b. Why did Scrooge look under the furniture in his room?

   c. Where did Scrooge think he saw Marley's face again?

   d. Why was Scrooge very surprised about the bell?

   e. Scrooge said, 'It's humbug! I don't believe it!' What was it?

   f. How did Scrooge know that the Ghost was Marley's?

   g. When did Marley make his chain?

   h. Marley's Ghost said, 'I must travel and never stop.' Why?

   i. What was the important thing that Marley's Ghost wanted to tell Scrooge?

   j. Why was Scrooge afraid when the window opened?

   k. What did the old ghost with the money-box want to do?

   l. Scrooge began to say 'Humbug!' Why do you think he stopped?

Subordinate clauses

'Because perhaps I ate a piece of meat or cheese and my stomach didn't digest it.'

This is a **subordinate clause**. Scrooge is saying *why* he doesn't believe in the Ghost, so it is a **subordinate clause of reason**. We usually find a subordinate clause with a **main clause** in full sentences. Look at the various types of subordinate clauses.

main clause	subordinate clause
Scrooge said this	**because** he didn't want to show his terror.
My business was goodness	**but** I didn't do anything good. (contrast)
Perhaps my stomach didn't digest some cheese	**so** you are only the consequence of a bad stomach. (result)
Marley made the chain	**when** he was alive. (time)
Sit down	**if** you can. (conditional)

2 **Match these main clauses with their subordinate clauses.**

Main Clause

a. ☐ The third Spirit will come

b. ☐ The old ghost was unhappy

c. ☐ In life Marley's spirit never left his office

d. ☐ Scrooge wasn't thinking about Marley

e. ☐ Scrooge was tired

f. ☐ At first the bell was quiet

g. ☐ The sound echoed around the house

h. ☐ I'll see hundreds of ghosts

i. ☐ You must see the three Spirits, Ebenezer

j. ☐ Scrooge was very frightened

Subordinate Clause

1. so now it must travel and never stop.

2. if you don't want to be like me.

3. when he put his key in the door.

4. because it couldn't help a poor woman and her baby.

5. but Scrooge wasn't frightened of echoes.

6. when the church bell strikes midnight.

7. but soon it made a very loud sound.

8. because the air was full of ghosts crying sadly.

9. so he went to bed and fell asleep immediately.

10. if I eat this candle.

Phrasal verbs

'Scrooge locked the door and put on his dressing-gown and slippers'.

In this sentence **put on** is a phrasal verb which means *'dressed himself'.* In English, many verbs can be followed by one or two prepositions (or adverbs) which can sometimes change their meaning.

Look at these examples.

*Did Marley's ghost really **come through** a locked door?*

*The ghost was **coming up** the stairs.*

3 Complete the following sentences using the phrasal verbs in the box, conjugated in the right tense.

> **look at   come into   go to   stay away from**
> **sit down   fall out of   take off   go up**

a. Scrooge lit a candle and then he .... the stairs slowly.

b. The thing .... Scrooge's room.

c. '.................................... – if you can,' said Scrooge.

d. Scrooge .... Marley's face for a moment, and then it was a knocker again.

e. 'When I was alive, I .... other people,' said the Spirit.

f. The Ghost .... the handkerchief and its chin dropped on its chest.

g. Scrooge .... his chair with horror.

h. Scrooge closed the window and .... the door. It was locked.

4 The Man who remembered Christmas.

For questions 1-12, read the text below and think of the word which best fits each gap. Use only one word in each gap. There is an example at the beginning (0).

A 1988 newspaper article claimed (**0**) *that* Charles Dickens invented Christmas. (**1**) .... is that possible? Isn't Christmas the ancient winter holiday to celebrate the birth of Jesus? Yes, of course it is. Then (**2**) .... did Dickens invent? Well, perhaps it (**3**) .... be better to say that he revived Christmas. *A Christmas Carol* was published (**4**) .... an important period in the history of England, and in the history of the world: England (**5**) .... becoming the first industrialised nation, and also the first nation in the history of the world with more people living in cities (**6**) .... in the countryside. Thousands and thousands of people (**7**) .... left the countryside to live and work in the cities. These people stopped practising the customs of their towns and provinces in the big city, and so the ancient ways of celebrating Christmas were gradually (**8**) .... lost. Also, in the new big cities there was no longer the time to celebrate the traditional Twelve Days of Christmas, which (**9**) .... from 25 December to 6 January. What Dickens did in *A Christmas Carol* and his other Christmas tales, was to remember the Christmases of his youth. He also helped the new city dwellers to remember all the joy and happiness of that great winter celebration, and to (**10**) .... alive the traditional Christmas Spirit. In fact, (**11**) .... a poor girl in London heard of Dickens' death in 1870 she said, 'Dickens dead? Then will Father Christmas die too?' There can be no (**12**) .... tribute.

Marley's Ghost

5 **Find words from the text to label this picture of Marley's ghost.**

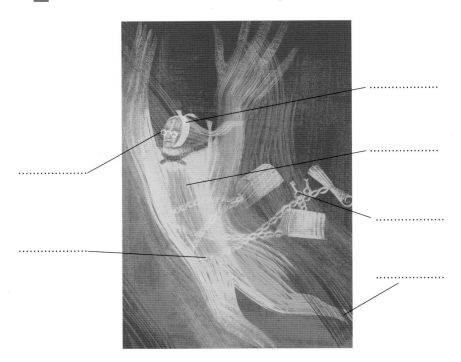

Now complete the sentences and check your answers in the text.

a. Marley's .... was transparent.

b. The chain was ....

c. The .... were heavy.

d. Marley's eyes were .... and ....

e. The .... was round Marley's head and ....

f. Marley's glasses ....

6 What happened at the following places in Scrooge's rooms? Write the sentence number (1-8) in the correct box.

A ☐ bed   E ☐ fire
B ☐ chair   F ☐ window
C ☐ door   G ☐ stair
D ☐ front door   H ☐ table

1. Scrooge looked under it.
2. The Ghost came up to Scrooge's room.
3. Scrooge saw Marley's face in the knocker.
4. The Ghost sat down.
5. The Ghost went out.
6. Scrooge thought he saw Marley's face again.
7. Scrooge fell asleep.
8. The Ghost came in.

Listening

track 03

7 Now listen to the first part of this chapter again and choose the correct answers. (A, B or C)

a. The Ghost
   A ☐ had a handkerchief in its hand.
   B ☐ had a chain round its body.
   C ☐ didn't have any clothes on.

b. Scrooge said he didn't believe it was a ghost because
   A ☐ he had a bad stomach.
   B ☐ he had eaten a candle.
   C ☐ he was very frightened.

c. When Marley was alive,
   A ☐ he loved money.
   B ☐ he wore a chain.
   C ☐ he helped people.

d. The Ghost

    A ☐ travelled to many places but it was a prisoner.

    B ☐ travelled faster than the wind.

    C ☐ never left prison when it was alive.

e. Scrooge looked frightened because

    A ☐ he didn't want to be like Marley.

    B ☐ he didn't want to see the ghosts.

    C ☐ the Ghost told him that he didn't have a chance.

f. How many spirits will come at one o'clock?

    A ☐ Two.

    B ☐ One.

    C ☐ Three.

Summary

8 **At the beginning of Chapter Two, Scrooge didn't believe in ghosts and he said, 'Humbug!' But at the end he stopped before saying it again. What happened to change his mind? Look at the sentences and put them in the correct order.**

a. ☐ All the bells in the house began to ring.

b. ☐ Marley's Ghost came into the room.

c. ☐ Scrooge saw Marley's face in the fire.

d. ☐ The Ghost shook its chains with a terrible cry.

e. ☐ Scrooge heard a noise like chains.

f. ☐ The Ghost said to Scrooge, 'You will see three Spirits.'

g. ☐ Scrooge saw Marley's face in the door knocker.

h. ☐ The Ghost took off the handkerchief.

i. ☐ The bell began to move.

j. ☐ Scrooge saw a lot of ghosts outside his window.

Some Christmas Ghosts

In northern countries like Great Britain, Christmas is traditionally a time for ghosts. People sit round a big fire and tell ghost stories.

In the sixteenth century two skulls [1] appeared every Christmas Day at Calgarth Hall in the English Lake District. Myles Phillipson, a magistrate, built Calgarth Hall on some land – but it wasn't his land. It belonged to a man called Kraster Cook, who lived in a cottage with his wife Dorothy. On Christmas Day Phillipson invited them to a big dinner and gave them a beautiful bowl [2] made of gold. Next day, soldiers arrested them. Phillipson was the magistrate at the trial. He said that the Cooks had taken his bowl. He sentenced them to death and they were hanged.[3] After Phillipson built Calgarth Hall, two skulls appeared on the stairs on Christmas Day. They returned every year on the same day. When the Phillipson family became poor and sold the Hall, the skulls disappeared.

For many years a ghost has returned to Sandringham in Norfolk, one of the historic homes of the British Royal Family. Every Christmas Eve, in the servants' rooms on the second floor, it throws Christmas cards into the air and pulls sheets from the beds.

There is a restaurant called Popjoy's next to the Theatre Royal in Bath, Avon. Just before Christmas, 1975, a man came in to have dinner. He went upstairs to the bar, bought a drink, looked at the menu and ordered his meal. While he was sitting on a sofa with his

1. **skulls** : craniums.
2. **bowl** : round container used for eating or for storing food.
3. **hanged** : suspended by the neck until dead.

drink a lady in old-fashioned [1] clothes came and sat next to him. Then she vanished. The man didn't stay for his meal. He ran downstairs, told his story to the waiter, and then ran out.

Sandringham in Norfolk.

1 Complete the tables with information from the texts, like the examples. For numbers 2 and 3 say what the ghosts did and try to imagine why the ghosts appeared! (Write your ideas in the 'Reason' section.)

1

Time Place	
People	
Ghost	two skulls appeared on the stairs
Reason	the spirits of Kraster and Dorothy wanted to frighten Phillipson, so that he wouldn't forget what he had done / said at the trial

---

1. **old-fashioned** : out of date.

2

Time **Place**	
People	
Ghost	
Reason	

3

Time **Place**	
Person	*a customer*
Ghost	
Reason	

2 **Answer these questions.**

a. Do you believe in ghosts?

Yes? How do you explain them? ....

No? Why not? ....

b. Would you like to see a ghost?

Why? ....

Why not? ....

Before you read

FCE **1** **Listen to the beginning of Chapter Three and complete the sentences.**

track 04

1. Scrooge went to bed at .... in the morning.

2. But when he woke up, the church clock ....

3. Scrooge asked himself whether Marley's ghost was
   ....

4. At one o'clock a hand .... of his bed.

5. The Spirit looked like a child, but also like ....

6. The Spirit had white hair and it .... clothes.

7. It told Scrooge that it was .... Christmas Past.

8. When Scrooge asked, 'Whose past?' it said, '.............................. .'

9. The Spirit said that it wanted ....

10. Then it said, 'Get up and ....'

Now read the text to check your answers.

Chapter Three

The First Spirit

track 04

Wh_en Scrooge woke up, it was very dark. The church clock struck twelve.

'Twelve!' said Scrooge, surprised. 'But it was after two o'clock when I went to bed. It's impossible! That clock is wrong.'

He got out of bed and went to the window, but he couldn't see much. It was dark, foggy and very cold. He went back to bed and began to think.

'Was it all a dream? Was Marley's Ghost really here?' he said to himself.

Suddenly he remembered the Ghost's words: 'The first Spirit will come at one o'clock tomorrow morning.' So he decided to wait and see. After a long time he heard the church clock.

'It's one o'clock!' said Scrooge. 'And there's nobody here!'

At that moment there was a great light in the room and the

curtains [1] of his bed opened. Yes, a hand opened the curtain in front of his face! He sat up and saw a strange person. It was small, like a child, but it was also like an old man. Its long hair was white but its face looked young. It was wearing white clothes with summer flowers on them. There was a piece of green holly [2] in its hand.

'Are you the first Spirit?' asked Scrooge.

'Yes, I am,' the visitor replied in a quiet voice.

'Who and what are you?'

'I am the Ghost of Christmas Past.'

'Whose past?'

'*Your* past.'

'Why are you here?'

'To help you.'

'I thank you,' Scrooge said. 'If you want to help me, let me sleep.'

'Get up and walk with me,' said the Spirit, and it took his arm.

Scrooge wanted to say that it was late, the weather was very cold, and his bed was warm. But the Spirit took him to the window.

'No, I'll fall!' Scrooge said.

The Spirit put its hand on his heart. 'If I touch you here, you won't fall,' it said.

Then they went through the wall, and suddenly they were standing on a road in the country. There was snow in the fields.

'Good Heavens!' [3] Scrooge cried. 'This is where I was born! I

---

1.   **curtains** [kɜːtənz] : (here) long pieces of material around the bed.
2.   **holly** : type of green plant, used as a decoration at Christmas.
3.   **Good Heavens!** ['hevənz] : exclamation, 'What a surprise!'

was a boy here!' And he remembered all his old feelings about the place.

'Your lip [1] is trembling,' said the Ghost. 'Are you crying?'

'No, no...' answered Scrooge. But a tear fell from his eye.

They walked along the road towards a little town with a bridge, a church and a river. Some boys came out of a school. They were laughing and singing because it was a holiday. They shouted 'Merry Christmas!' to each other.

'They are all in the past,' the Ghost said. 'They are only shadows.' [2]

Scrooge knew all of them and he felt suddenly happy. Why did his cold eyes and heart become warm with joy? What did merry Christmas mean to him? He didn't like Christmas!

'The school is not empty,' said the Spirit. 'One child is still there. He hasn't got any friends.'

'I know, I know,' Scrooge said. And there were big tears in his eyes.

They went into the school, a big, old, dark place. Inside there was a long classroom. It looked sad and empty, with only a few desks and chairs in it. A little boy was sitting at one of the desks. He was reading a book by a small fire. Scrooge sat down on a chair and cried because he knew that the little boy was himself many years ago.

'That's me,' he said. 'I was left here one Christmas. Poor boy! Oh, I would like to... but it's too late now!'

'What is it?' asked the Spirit.

1. **lip** :

2. **shadows** : (here) spirits, ghosts.

'Nothing. You see, there was a poor boy outside my office last night. He was singing a Christmas carol. But I didn't give him anything and I told him to go away.'

The Spirit smiled. 'Let's see another Christmas!'

Then everything changed. The boy was bigger, and the room looked older and darker. Scrooge saw himself again. He was walking sadly up and down. Then the door opened and a little girl ran in. She was younger than the boy.

'Dear, dear brother!' she said happily. And she put her arms round his neck and kissed him. 'I've come to bring you home – home, home!'

'Home, Fanny?' the boy asked.

'Yes! Home for ever and ever!' [1] the girl laughed. 'Father is kinder [2] now and he wants you to come home. He sent me in a coach [3] to fetch you. [4] Oh, you'll never come back to this horrible school! And we'll be together for Christmas! I'm so happy!'

She began to pull him towards the door.

'Bring Master Scrooge's luggage [5] to the coach!' somebody shouted in a terrible voice.

It was the teacher, and when he came in, the boy was very frightened.

'Goodbye, Master Scrooge!' said the teacher in his terrible voice.

'Goodbye, sir,' the boy answered, trembling.

But when he got into the coach with his sister, he felt happy.

---

1. **for ever and ever** : for always.
2. **kinder** ['kaɪndə] : more understanding.
3. **coach** : (here) covered, wheeled transport pulled by a horse.
4. **fetch you** : take you back home.
5. **luggage** ['lʌgɪdʒ] : bags, suitcases etc. for travelling.

A Christmas Carol

'Your sister had a very good heart,' said the Ghost. 'When she died, she left one child – your nephew.'

'Yes.' Scrooge remembered the conversation with his nephew in his office the afternoon before, and he felt bad about it.

Suddenly they were standing at the door of an office in the city. It was Christmas again.

'I know this place very well! And there's old Mr Fezziwig – alive again! Oh, dear old Fezziwig!'

Mr Fezziwig was a fat, happy man with a red face. He was working at a desk.

'Hey! Ebenezer! Dick!' he shouted. 'Stop your work!'

Scrooge, now a young man, came in with his friend Dick.

'It's Christmas Eve, boys! We must celebrate!' said Fezziwig. 'Let's stop work and close the office.'

So they put away all the books and papers and made a big fire. Then a man came in and started to play the violin. Mrs Fezziwig and the three Miss Fezziwigs arrived, and then a lot of young people came, and everybody began to dance to the music. Then there were games and more dances; cake and hot wine and more dances. And there was lots of roast beef and beer, and mince pies [1] too. It was a wonderful party. At eleven o'clock everybody said 'Merry Christmas!' and the party finished. While Scrooge was watching all this, he laughed and sang and wanted to dance. He remembered it all and enjoyed it very much.

'You and Dick and everybody loved Mr Fezziwig,' the Ghost said to him. 'But why? That party was a very small thing. It cost only three or four pounds. So why did you all love him so much?'

---

1. **mince pies** [mɪns'paɪz] : special Christmas sweet made of pastry, with dried fruit inside.

'A small thing!' answered Scrooge. 'No! Fezziwig was our manager, so he could make us happy or unhappy. He could make our work easy or hard. He gave us a lot of happiness – and that was like a fortune in money!'

Then Scrooge looked sadly at the Ghost.

'What are you thinking about?' it asked.

'I... was thinking that I would like to speak to my clerk now...'

'Come, there isn't much time,' said the Ghost. 'We must be quick.'

At that moment the scene vanished and they were standing in the open air. Scrooge saw a man of about forty. It was himself again, and his face showed the first signs [1] of the problems of business and a passion for money. He was sitting next to a young girl dressed in black. It was his fiancée [2] Belle. She was crying quietly.

'You love something more than me, Ebenezer,' she said.

'Oh? What?'

'Money. You are afraid of life, you are afraid of the world, and so you do only one thing: make money. Then you feel more secure. Money is your passion now.'

'No,' he said angrily. 'My feelings for you haven't changed, Belle!'

'But you *have* changed. When you promised to marry me, you were a different person.'

'I was a boy,' he said.

'And so my love is nothing to you now. You aren't happy with me and you don't want to marry me.'

'I've never said that.'

'Not in words, no – but I know it's true. I haven't got any

1. **signs** [saɪnz] : indications.
2. **fiancée** : girl with whom you are engaged to be married.

money so you don't want me. Well, you're free to go. I hope you will be happy.' And Belle went sadly away.

'Spirit!' Scrooge cried. 'Don't show me any more! Take me home!'

'There's one more scene.'

'No! No more! I don't want to see it!'

But suddenly they were in a room where a beautiful young girl was sitting near a big fire. Next to her sat her mother. This was Belle, now older. The room was full of children and there was a lot of noise. But Belle and her daughter liked it, and the daughter began to play with the children. Then the father came in with a lot of Christmas presents. He gave them to the children and they laughed and shouted happily. Finally, they went to bed and the house was quiet. The father sat by the fire with his wife and daughter. Scrooge looked at them and thought: 'How sad that *I* don't have a wife and daughter!'

'Belle,' said the husband to his wife. 'I saw your old friend this afternoon.'

'Who was it? Mr Scrooge?'

'Yes. I passed his office window and he was there. He hasn't got a friend in the world. His old partner Marley is dying.' [1]

'Spirit, take me away!' said Scrooge.

'These things happened,' the Ghost answered, 'and they cannot be changed.'

'Please take me back! I can't watch this any more!'

At that moment the Spirit disappeared and Scrooge was in his bedroom again. He felt very tired, so he got into bed and fell asleep.

---

1. **dying** : present participle of the verb 'to die'.

Summary

1 **Match the two parts of the sentences to make a summary of Chapter Three. The first is done for you.**

a. When Scrooge woke up
b. At one o'clock
c. It took Scrooge
d. and he saw himself
e. It was Christmas, but
f. Then he saw himself again
g. She said
h. Suddenly the Ghost showed him
i. His kind, old manager Mr Fezziwig
j. Scrooge thought sadly
k. Then he saw himself at 40
l. She was crying because
m. She told him
n. Suddenly Scrooge saw her
o. She was married
p. Her husband told her
q. Scrooge felt very sad
r. Then he was in his bedroom again

1. ☐ when he was a boy.
2. ☐ that he could come home with her.
3. ☐ she knew that Scrooge didn't want to marry her.
4. ☐ when she was older.
5. ☐ about his clerk Bob Cratchit.
6. ☐ so he got into bed and fell asleep.
7. ☐ into the past
8. ☐ that now he loved money more than her.
9. ☐ and had a big, happy family.

10. ☐ with his fiancée Belle.

11. [a] it was midnight.

12. ☐ the office where he began his first job.

13. ☐ he was all alone at school.

14. ☐ and he asked the Ghost to take him away.

15. ☐ decided to give a Christmas party.

16. ☐ that Marley was dying and Scrooge was all alone.

17. ☐ the Ghost of Christmas Past arrived.

18. ☐ with his sister Fanny.

2 A journalist is interviewing Scrooge for the *Morning Gazette* newspaper. Using information from Chapter Three, write Scrooge's answers. (One is done for you.)

Journalist: Where were you born, Mr Scrooge?

Scrooge: 0 In the country near a small town ....

Journalist: Do you have any brothers or sisters?

Scrooge: 1 ....

Journalist: What was her name?

Scrooge: 2 ....

Journalist: Was she older than you?

Scrooge: 3 ....

Journalist: Was she a good sister?

Scrooge: 4 ....

Journalist: Were you happy at school?

Scrooge: 5 ....

Journalist: Why not?

Scrooge: 6 ....

Journalist: Where was your first job and who was your boss?

Scrooge: 7 ....

Journalist: What was he like?

Scrooge: 8 ....

Journalist: Did you have a fiancée when you were younger?

Scrooge: 9 ....

Journalist: Belle! That's a lovely name! What happened to her?

Scrooge: 10 ....

 3 **You are going to read a summary of the story so far. Choose from the list A-I the sentence which best fits each space (1-7). There is an example at the beginning (0).**

Ebenezer Scrooge was a mean old businessman. **(0)** *H*. Nobody liked him, nobody talked to him, but he didn't care. He was a cold and solitary man. **(1)** ..... . He wanted to say Merry Christmas and invite Scrooge to dinner on Christmas Day. **(2)** ..... . Then two fat men came in and asked Scrooge to give some money for the poor. But he refused and they went away. When it was time to go home he didn't want to give his clerk, Bob Cratchit, a day's holiday. **(3)** ..... .

At his house Scrooge saw the face of his dead partner Marley in the door knocker. **(4)** ..... . The ghost wore a long, heavy chain, which it had made for itself in life. When Marley was alive he stayed away from people, locked himself in his office, and just made money. **(5)** ..... . But he told Scrooge that there was still a chance for him to escape the same terrible consequences with the help of three Spirits. **(6)** ..... . He saw himself as a sad, lonely, frightened boy. Then the Spirit showed him more scenes from his life. **(7)** ..... . He didn't marry her but chose a life of making money instead of happiness.

A He thought only about money.

B But Scrooge said Christmas was a humbug and refused.

C So now he had no rest or peace, and he was very unhappy.

D The first Spirit was the Ghost of Christmas Past and it took Scrooge to his old school.

E The Ghost left through the window.

F One Christmas Eve Scrooge's nephew Fred arrived at his uncle's office.

G He also told him to arrive extra early on the 26th.

H One was with his fiancée Belle.

I Later, Marley's ghost appeared to Scrooge.

4 **For questions 1-10, complete the second sentence so that it has a
similar meaning to the first sentence, using the word given. Do not
change the word given. You must use between two and five words,
including the word given. There is an example at the beginning (0).**

0. It was dark and foggy, so Scrooge couldn't see much.
   because
   Scrooge couldn't see much *because it was dark*.. and foggy.

1. 'Was it a dream?' Scrooge asked himself.
   been
   Scrooge asked himself .... a dream.

2. The Spirit looked like a child, but also like an old man.
   both
   The Spirit looked like .... an old man.

3. 'It's one o'clock and there's nobody here!', Scrooge said.
   isn't
   'It's one o'clock and there .... here!' Scrooge said.

4. After a long time Scrooge heard the church clock.
   before
   It was .... Scrooge heard the church clock.

5. They walked towards a little town with a bridge, a church, and a
   river.
   which
   They walked towards a little town .... a bridge,
   a church, and a river.

6. 'If I touch you here, you won't fall,' said the Spirit.
   will
   'You .... I touch you here,' said the Spirit.

7. When he saw himself as a boy Scrooge was too sad to speak.
   that
   When he saw himself as a boy Scrooge was ....
   not speak.

8. Fanny was younger than her brother.
   wasn't
   Fanny .... her brother.

9. 'Spirit, don't show me any more!' Scrooge cried.
to
'Spirit, I don't want .... me any more!'
Scrooge cried.

10. Scrooge got into bed and fell asleep because he felt very tired.
so
Scrooge felt very tired .... and fell asleep.

Before you read

1 Which of these pictures do you associate most with Christmas? Which do you think are the most important. And the least? Discuss with your partner.

2 Now select one of the following questions and talk about it with your partner.

- Can you think of anything else you associate with Christmas?
- Do you think that we celebrate Christmas in the right spirit today?
- Do you know what a 'White Christmas' is and have you ever seen one?
- Do you think that it's possible to have the Christmas spirit every day of the year?
- Do you think Christmas would be better or worse without TV?

The Christmas Story

In the pagan world there was a midwinter festival after December 22nd – the shortest day of the year – when the days become longer. The Celts built big fires that symbolised the return of the sun. The Romans had the Saturnalia when people ate, drank and danced. Then, in the fourth century AD, Pope Julius I made December 25th the official date of Christ's birth. So Christmas combines a pagan religious festival and a Christian festival.

Ancient customs became a part of Christmas: for example, decorating houses with mistletoe [1] or holly. [2] From the pagan Yule

Glad Tidings (end of the 19th century) by William M. Spittle.

1. **mistletoe** : plant traditionally associated with Christmas.
2. **holly** : traditional Christmas plant.

festival of the sun in Scandinavia we get Yule logs and Yule candles at Christmas, which symbolise fire and light. Later, Saint Nicholas, the patron of children, became associated with Christmas. He was famous for his generosity, and people said he often left presents for them. The tradition of carol-singing began many centuries ago.

But some traditions are quite modern. The image of a fat, happy Father Christmas in red and white clothes came from America at the end of the 19th century. Christmas crackers [1] first appeared in the 1870s. In the 1840s Queen Victoria and her husband Albert introduced the custom of decorating a

Christmas, cover of the *Saturday Evening Post* (1926) by Norman Rockwell.

Christmas tree. They used candles, not lights, and they put children's toys and presents on the tree. They had Christmas stockings [2] too. People also started sending Christmas cards in the 1840s. Other traditions go back to earlier centuries. Christmas pudding [3] was first made about 1670. Before the 16th century there was no turkey at Christmas. People ate goose and other birds.

1. **crackers** : paper tube with toys, sweets etc. inside.

2. **stockings** :

3. **Christmas pudding** : traditional dessert.

In Dickens's time, people usually gave presents at the New Year. And Boxing Day (December 26th) was not a holiday. St. Stephen's Day is called Boxing Day in Britain – but not because the British celebrate boxing on that day! Today a 'Christmas box' is any type of present. But years ago it was the custom to give servants a 'Christmas box' – a present of money or clothes – on December 26th. This custom originated with the early Christian Church when the priests broke the money boxes on the day after Christmas and gave the contents to poor people.

1 **Which of the following statements are true (T) or false (F)? Correct the false ones.**

	T	F
a. There was a pagan midwinter festival before the shortest day of the year.	☐	☐
b. The early Church changed the pagan festival into a Christian one.	☐	☐
c. When he became associated with Christmas Saint Nicholas was already the patron of children.	☐	☐
d. The modern Father Christmas was a European idea.	☐	☐
e. Christmas trees and Christmas cards began at about the same time.	☐	☐
f. Christmas pudding and turkey appeared about the same time.	☐	☐
g. In Dickens's time, people had two or three days' holiday at Christmas.	☐	☐
h. Boxing Day has no connection with the sport of boxing.	☐	☐
i. A modern-day 'Christmas box' is a box containing money.	☐	☐
j. Boxing Day originated with a religious custom of giving presents to servants.	☐	☐

Writing

2 Imagine that you are a British girl writing a letter to your pen friend who lives in India where Christmas is not an important festival. Read this letter from her, on which you have made some notes. Then, using all the information in your notes, write a suitable reply.

> *Dear Mary, thank you for your letter. Here in India we celebrated Diwali, our festival of lights at the end of Autumn.* — Christmas Dec. 25th
>
> *We light small lamps called diye and put them in every corner of the house. They welcome the Hindu god Rama.* — Father Christmas
>
> *Diwali lasts for two days. We eat small balls of rice called khil and we give Diwali cards and Indian sweets to our friends.* — turkey, mince pies, Christmas pudding
>
> *Write and tell me how you celebrate Christmas in Britain. Do you send cards and give sweets to your friends?* — Yes, we do.
>
> No, we don't
>
> *Your friend*
> *Suman*

Write a letter of between 120 and 150 words in an appropriate style.

60

Before you read

FCE **1** **Listen to the beginning of Chapter Four. For questions 1-7 choose the best answer A, B, or C.**

track 05

1. Why did Scrooge get out of bed?
   - **A** ☐ A voice called his name.
   - **B** ☐ He saw a strong light in the next room.
   - **C** ☐ It was one o'clock.

2. What was different about Scrooge's room?
   - **A** ☐ There was a sofa and a fireplace in it.
   - **B** ☐ It was lit by a giant ghost with a torch.
   - **C** ☐ It was full of good Christmas things.

3. When Scrooge was in front of the Spirit, it told him
   - **A** ☐ to look at it.
   - **B** ☐ to come in.
   - **C** ☐ not to be frightened.

4. It was the first time Scrooge had seen
   - **A** ☐ a ghost.
   - **B** ☐ the Ghost of Christmas Present.
   - **C** ☐ a kind, happy person.

5. The Spirit's brothers were
   - **A** ☐ The Ghosts of all the Christmases in the past.
   - **B** ☐ all younger than the Spirit.
   - **C** ☐ more than eighteen hundred years old.

6. Where did the Spirit take Scrooge first?
   - **A** ☐ To a busy city street.
   - **B** ☐ To a baker's.
   - **C** ☐ To the suburbs.

7. When the Spirit touched the dinners with its torch
   - **A** ☐ they cooked extra quickly.
   - **B** ☐ they became even better.
   - **C** ☐ the people became happy.

Chapter Four

The Second Spirit

crooge woke up, opened his bed-curtain and looked around. He was ready to see anything, but when one o'clock struck, nothing happened. After a while he saw a strong light in the next room. He got out of bed and went slowly to the door.

'Scrooge!' said a voice. 'Come in, Ebenezer!'

The room was his room, but it was different. On the walls there was some green holly with red berries, and mistletoe and ivy. [1] In the fireplace was a great fire. On the floor there was a lot of food: turkey, goose, [2] chicken, rabbit, pork and sausages, as well as mince pies, puddings, fruit, cakes, and hot punch. [3] And

---

1. **ivy** ['aɪvi] : evergreen climbing plant.

2. **goose** :

3. **punch** [pʌntʃ] : (here) alcoholic drink made with wine mixed with fruit or fruit juice.

on the sofa sat a very large man – a giant – and he was holding up a torch. [1]

'Come in!' said the Ghost.

Scrooge went and stood in front of this giant, but he didn't look at it. He was too frightened.

'I am the Ghost of Christmas Present,' said the Spirit. 'Look at me!'

So Scrooge looked. He saw that the Spirit was smiling. It had kind, gentle eyes. There was holly round its long dark hair. Its face was young and happy.

'You have never seen anybody like me before,' it said.

'Never.'

'And you have never met any of my brothers?'

'No. How many brothers have you got?'

'More than eighteen hundred. I am the youngest.'

'Spirit,' Scrooge said, 'take me where you want. I learnt a good lesson last night.'

'Touch my clothes!'

When Scrooge did this, the room disappeared and he stood in the city streets on Christmas morning. There was a lot of snow. Some people were playing and throwing snowballs. Others were buying food in the shops. It was a busy, cheerful [2] scene, and the bells were ringing.

Then a lot of poor people came along the street with their Christmas dinners of goose or chicken. They were taking them to the baker's shops to be cooked in the oven. [3] The spirit took

---

1. **torch** : portable light (here) a flame.
2. **cheerful** : happy.
3. **baker's shops ... oven** [ˈʌən] : this was an old custom. Poor people often didn't have ovens.

A Christmas Carol

Scrooge to one of these shops and touched some of the dinners with its torch.

'What are you doing?' Scrooge asked.

'I am making these dinners extra good so the people will be happier,' it replied, smiling.

(3) After a while Scrooge followed the Ghost to the suburbs [1] of the city. They went to the house of Bob Cratchit, his clerk. The kind Ghost touched the house with its torch. Then they went in. Mrs Cratchit and her daughter Belinda were preparing the table for Christmas dinner. Young Peter Cratchit was helping them. Suddenly two little Cratchits ran in and shouted that the goose was ready at the baker's. Then the oldest daughter Martha arrived, and after her came Bob with his little son Tiny Tim on his shoulder. The child was a cripple [2] and he walked around on a small crutch. [3]

(4) Young Peter went to fetch the goose. When he came back, all the children in the family shouted 'Hurray!' because they didn't often eat goose. Belinda made some apple sauce; Mrs Cratchit prepared the potatoes and the gravy; [4] Martha put the hot plates on the table. Finally, everything was ready. When Mrs Cratchit cut the goose, everybody cried 'Hurray!' again, and Tiny Tim hit the table with his knife. The goose was small, but they all said it was the best goose in the world and ate every bit of it. [5] Then Mrs Cratchit brought in the Christmas pudding with brandy on it. She

---

1. **suburbs** : areas on the outside of a town or city.
2. **cripple** : someone with a physical disability.
3. **crutch** :
4. **gravy** ['greɪvi] : brown sauce made from the juices of meat.
5. **every bit of it** : all of it.

lit the brandy with a match, and when they were all eating, they said, 'Oh, what a wonderful pudding!' Nobody said or thought that it was a very small pudding for a big family.

After dinner the Cratchits sat round the fire. They ate apples and oranges, and hot chestnuts. [1] Then Bob served some hot wine.

'A Merry Christmas to us all!' he said.

'A Merry Christmas!' the family shouted.

'And God bless everyone!' said Tiny Tim in his weak [2] voice.

He sat very near his father. Bob loved his son very much and he held Tiny Tim's thin little hand.

'Will Tiny Tim live, Spirit?' Scrooge asked.

'I see an empty chair,' replied the Ghost, 'and a small crutch. But not Tiny Tim. If the future does not change, the child will die.'

'No, no!' said Scrooge. 'Say he will live, kind Spirit!'

'If the future is not changed, he will not see another Christmas. But you think that's a good thing, don't you? You said there are too many people in the world.'

Scrooge didn't answer and he didn't look in the Ghost's eyes. He felt very bad.

'Those were wicked [3] words, Ebenezer Scrooge,' the Ghost continued. 'Do you think you can decide who will live or die? Are you better than this poor man's child, or millions like him? Perhaps you are worse in God's eyes!'

Scrooge trembled and looked at the ground. Suddenly he heard his name.

---

1. **chestnuts** :

2. **weak** : not strong.
3. **wicked** ['wɪkɪd] : malevolent, very bad.

A Christmas Carol

6 'Mr Scrooge! Let's drink to Mr Scrooge!' It was Bob Cratchit and he was holding up his glass.

'Drink to Mr Scrooge!' said Mrs Cratchit angrily. 'Drink to that hard old miser! What are you saying, Robert Cratchit?'

'My dear – the children. It's Christmas Day.'

'I know that, but I would like to tell Mr Scrooge what I think of him! You know how bad he is.'

'My dear, it's Christmas Day.'

'Well, I'll drink to him because it's Christmas. A Merry Christmas and a Happy New Year, Mr Scrooge! – but you won't be merry or happy, I know.'

The children drank to Scrooge too, but his name was like a dark shadow in the room and for a few minutes they were silent. Then they told stories and sang songs, and they felt better. The Cratchits were poor and they looked poor. Their clothes were old; there were holes in their shoes. Bob Cratchit's salary [1] was very small. He never had enough money and there was never much food in the house. But the family was contented now because it was Christmas. Scrooge watched them carefully. He listened to them well. And he looked at Tiny Tim very often before the family scene vanished.

7 It was dark now, and snow was falling. Scrooge and the Ghost walked along the streets and saw great fires in the houses, where families and friends were enjoying Christmas together. The Ghost was happy to see the celebrations. It laughed, and where it passed, people laughed too. And then Scrooge heard a loud, happy laugh. It was his nephew's. He saw him in a bright, warm

---

1. **salary** : money earned each month.

room. When his nephew laughed, the other people in the room laughed with him.

'He said that Christmas was a humbug!' the nephew laughed. 'And he believed it too!'

'He's stupid and bad, Fred,' said his wife.

'Well, he's a strange man, and he isn't very happy.'

'But he's very rich, Fred.'

'Yes, my dear, but he doesn't do anything with his money. He doesn't help others, and he lives like a poor man.'

'Nobody likes him. I don't like him. He makes me angry.'

'I'm not angry with him. I feel sorry for him because he doesn't enjoy his life. He never laughs. He didn't want to eat with us today, but I'm going to ask him every year. I'll say, '"How are you, Uncle Scrooge? Come and eat with us."'

Then they played some music and sang. After that, there were games. When they played twenty questions,[1] Scrooge forgot that they couldn't hear him and he shouted his answers. Then his nephew thought of something and everybody asked him questions.

'Is it an animal?'

'Yes.'

'Does it live in the city?'

'Yes.'

'Is it a horse?'

'No.'

It wasn't a dog, a cat or a pigeon. It made horrible noises, sometimes it talked, and nobody liked it.

---

1. **twenty questions** : a game. Someone thinks of a person/an object; other people ask (a maximum of 20) questions to discover who/what it is.

'I know what it is!' shouted Fred's wife. 'It's your Uncle Scro-o-o-o-oge!'

She was right.

'A Merry Christmas and a Happy New Year to the old man!' said Fred.

Scrooge wanted to say this to Fred, but the scene vanished and he and the Ghost travelled again. Scrooge noticed that the Spirit looked older. Its hair was grey now.

(9)   'Is your life so short?' he asked.

'Very short. It ends tonight at midnight. It's eleven forty-five. I haven't got much time. Look – look down here!'

The Spirit opened its coat and Scrooge saw two children on the ground, a boy and a girl. They were very thin. Their clothes were old and poor, and they were trembling with cold. They looked very hungry. Their eyes were sad. They looked older than children and they were ugly, ¹ like monsters. Scrooge was shocked.

'Are they yours?' he asked.

'No. They are Man's. They belong to humanity.'

'Haven't they got a house or a family?'

'Aren't there a lot of prisons?' the Spirit replied. 'And aren't there any workhouses?'

'Oh, no – no! Those are my words!' Scrooge cried.

The church clock struck twelve. He looked around for the Ghost but it wasn't there. Then he remembered old Jacob Marley's words:

'The third Spirit will come at twelve midnight.'

---

1. **ugly** : not beautiful.

Summary

1 **Chapter Four has been divided into 9 parts. Choose from the list A-J the sentence which best summarises each part (1-9). There is one extra sentence which you do not need to use.**

A ☐ Meet the Cratchits.

B ☐ What's the right answer?

C ☐ Out for a lesson with the youngest brother.

D ☐ Poor – but rich in the Christmas spirit!

E ☐ There's a giant ghost in my room!

F ☐ The shocking truth.

G ☐ They're talking about me.

H ☐ Is a tragedy waiting in the future?

I ☐ A magic torch.

J ☐ Big family, small dinner.

Reported Speech

'He said that Christmas was a humbug.'

Fred is telling other people what his uncle said, so this is a **reported statement.**

Look at these sentences.

Direct Statement	Reported Statement
'I **am** the Ghost of Christmas Past,' said the Spirit.	The Spirit said (that) it **was** the Ghost of Christmas Present.
'I **don't like** Scrooge,' Fred's wife said.	Fred's wife said (that) she **didn't like** Scrooge.
'I **learnt** a good lesson last night,' said Scrooge.	Scrooge said (that) he **had learnt** a good lesson the night before.

We can see that the tense of the verbs move back one tense in each example, Present Simple becomes Past Simple etc.

We can also report direct questions, requests, and commands. Look at these examples:

Direct	Reported
(Question) 'What **are** you **doing**?' Scrooge asked the Spirit.	Scrooge asked the Spirit what it **was doing**.
(Request) 'Can/Would you go and fetch the goose, Peter?' Bob asked.	Bob asked Peter to go and fetch the goose.
(Command) '**Come in**!' said the Spirit.	The Spirit told Scrooge **to come in**.
(Question) '**Will** Tiny Tim live, Spirit?' Scrooge asked the Spirit.	Scrooge asked the Spirit if Tiny Tim **would live**.

2 **Change these *direct statements, questions, requests,* and *commands* into reported speech.**

a. 'Nobody likes Scrooge,' said Fred's wife.

b. 'Touch my clothes!' the Spirit said to Scrooge.

c. 'I haven't got much time,' said the Spirit.

d. 'Would you drink to Scrooge, my dear?' Bob asked his wife.

e. 'Is it a horse?' Fred's wife asked.

f. 'What are you saying?' Mrs Cratchit asked her husband.

g. 'Will Tiny Tim see another Christmas, Spirit?' Scrooge asked.

h. 'Look at me!' the Ghost said to Scrooge.

3 **Now change these *reported statements, questions, requests,* and *commands* into direct speech.**

a. Mrs Cratchit asked Peter if he could help her to prepare the table.

b. Bob Cratchit said that it was Christmas Day.

c. Scrooge asked the Spirit if its life was short.

d. Mrs Cratchit said she would drink to Scrooge because it was Christmas.

e. Fred's wife asked if the animal lived in the city.

f. The Spirit told Scrooge to look down on the ground.

g. Mrs Cratchit asked Belinda to put the plates on the table.

h. Scrooge said that there were too many people in the world.

Who did what?

4 **A Who in Chapter Four**

1. carried Tiny Tim on his shoulder?
2. put hot plates on the table?
3. shouted, 'Hurray!'?
4. said, 'Say he will live, kind Spirit!'?
5. said, 'My dear, it's Christmas Day.'?
6. shouted, 'It's your Uncle Scro-o-o-o-oge!'?
7. said, 'I haven't got much time.'?
8. was shocked?

B If possible, write why they said or did these things.

1. ....
2. ....
3. Because they didn't eat goose very often. ....
4. ....
5. ....
6. ....
7. ....
8. ....

Check your answers in the text.

Present Perfect

'You have never seen anybody like me before.'

The Present Perfect is formed with the present tense of **have** or **has + the past participle**. It is a combination of the past and the present, and often provides a strong connection between the two. We can use it to talk about a period of time that began in the past and continues up to the present.

So the Spirit means that Scrooge hasn't seen anybody like it from an indefinite time in the past **until now**. This period covers Scrooge's life up to the present moment.

Never is used in statements, **ever** in questions.

*I have **never** seen a ghost.* (in my life)

*Have you **ever** seen a ghost?* (in your life)

5 **Write sentences using *never* and *ever* with the following words.**

    a. You/meet/any of my brothers?

    b. Bob Cratchit/earn/much money.

    c. Tiny Tim/walk/without a crutch.

    d. Fred/eat/turkey?

    e. You/have/a white Christmas?

    f. Scrooge/laugh/in his life.

    g. Fred's wife/like/Scrooge.

    h. The Cratchit children/play/twenty questions?

'Who am I?'

6 **A** **You are going to play a game like twenty questions.**
Read the clues and write the name of a character in the story.

I'm kind and happy	I make people happy
I'm big	I work in an office
I've got hundreds of brothers	I have three daughters
My life is short	I am fat
Who am I? ...	Who am I? ....

    B **Now *you* write the clues for 2 people in the story and play 'Who am I?' with your partner.**

....

....

....

Before you read

track 06

1 Listen to the beginning of Chapter Five and decide whether the statements are true (T) or false (F). Then check your answers in the text.

	T	F
a. Scrooge couldn't see the phantom because it was invisible.	☐	☐
b. The phantom said it was the Ghost of Christmas Yet To Come.	☐	☐
c. In answer to Scrooge's question the Ghost only pointed with its finger.	☐	☐
d. At first Scrooge couldn't follow the Ghost because he was frightened.	☐	☐
e. Scrooge went with the Ghost because it didn't speak to him.	☐	☐
f. The Ghost took Scrooge to a place where businessmen made money.	☐	☐
g. Time and money were very important to the businessmen.	☐	☐
h. The three men were talking about somebody's death.	☐	☐
i. They were sad because they liked the dead person.	☐	☐
j. Scrooge knew they were talking about him.	☐	☐

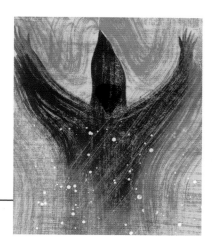

Chapter Five

The Last of the Spirits

track 06

Another phantom was coming towards him. It was tall and silent. Scrooge couldn't see its face or its body because it was wearing long, black clothes and a black hood. [1] There was something mysterious about it. When it came near him, Scrooge was very frightened. It didn't speak or move.

'Are you the Ghost of Christmas Yet To Come?' [2] he asked.

The Spirit didn't answer, but its long, white hand came out from the black clothes and pointed [3] down.

'Are you going to show me things from the future?' Scrooge asked.

His legs were trembling a lot and so he couldn't follow the Ghost when it moved away. It stopped and waited for him. He

---

1. **hood** [hʊd] : part of a jacket or coat that covers the head.
2. **Yet To Come** : in the future.
3. **pointed** : indicated with its finger.

couldn't see its eyes but he felt that they were looking at him. This Ghost was the most frightening of the three.

'Ghost of the Future!' he cried. 'I'm very frightened of you! But I know that you want to help me so I'll go with you. Please speak to me!'

It made no reply. Its long hand pointed ahead. [1]

'All right, I'll come,' said Scrooge.

So the Ghost carried him to the centre of London. At a place called the Exchange [2] he saw a lot of businessmen. Their pockets were full of money. They were walking around and talking to each other. They often jingled [3] the money in their pockets and looked at their watches. Scrooge knew many of them. When the Ghost stopped near three men, he could hear their conversation.

'No, I don't know much,' said one very fat man. 'I only know he's dead.'

'When did he die?' another man asked.

'Last night, I think.'

'I thought he would never die. Was he very ill?'

'God knows.' [4]

'What about his money?' asked a man with a very red face.

'I don't know,' replied the fat man. 'He hasn't left it to me.'

Everybody laughed.

'The funeral will be very cheap because only a few people will go,' the fat man continued.

---

1. **ahead** [ə'hed] : in front.
2. **the Exchange** : stock exchange, where financial transactions take place.
3. **jingled** ['dʒɪŋgld] : made a soft sound like small bells.
4. **'God knows'** : I don't know.

A Christmas Carol

'I'll go if there's a big lunch,' the red-faced man said.

Another laugh. And then the men went away. Scrooge looked at the Ghost.

'Who are they talking about?' he asked.

But the Ghost said nothing. It went into the street and showed Scrooge two men. He knew them. They were rich and important businessmen. First they said hallo. Then one of them said:

'Well, he's finally dead.'

'Yes, I've heard,' answered his companion. 'Cold, isn't it?'

'Very cold. But it's the right weather for Christmas. Would you like to come ice-skating?'

'No, thank you. I'm too busy. Good morning.'

That was the end of their conversation. Scrooge was surprised. Who were they talking about? He couldn't think of anybody. Old Marley died seven years ago and this Ghost was showing him the future. He decided to wait and see. He looked around but couldn't see himself anywhere. Wasn't he there in the future? Silent and black, the Ghost stood near him. He knew that it was watching him and he trembled.

end

They went into a poor part of the city where the streets were dirty and narrow. 1 There were dark shops and houses, and the people looked ugly and miserable. 2 A lot of them were drunk. 3 Rubbish 4 was everywhere, and there were bad smells. 5 The quarter was full of dangerous criminals. Scrooge followed the

---

1. **narrow** : not wide.
2. **miserable** : very sad.
3. **drunk** : inebriated with alcohol.
4. **Rubbish** : things that people throw away.
5. **smells** : odours.

Ghost into a small, dark shop. It was full of dirty, old things –
bottles, clothes, keys, chains. A man of about seventy with grey
hair sat near a fire and smoked his pipe. Then a woman came in
with a big, heavy box in her arms. She put it on the floor and sat
down.

'Open it, Old Joe,' she said, 'and give me the money.'

The man opened the box. 'What are these?' he said. 'Bed-
curtains! Did you take them while he was in bed?'

'Yes. Why not? There was nobody with him. There are
blankets [1] too.'

'His blankets?'

'Of course! He won't need them where he's going. Here's a
beautiful, expensive shirt too. He was wearing it for his funeral. I
thought, "What a pity! This is a very fine shirt but nobody will
wear it again." So, I took it off him.'

'You did well, madam,' laughed Old Joe. 'You're a clever
woman and you'll make a fortune one day.'

'I must think of myself, like him. He was a selfish [2] old miser. I
cleaned his rooms and his clothes. I worked very hard for him but
he never gave me anything. I wanted to take more things but his
housekeeper [3] took them before me.'

Just then the housekeeper came in. She had a large bag full of
sheets, towels, [4] clothes, and shoes.

---

1. **blankets** : large pieces of material (made of wool) to keep someone warm in bed.
2. **selfish** : egoistic. He thought too much about himself and what he wanted.
3. **housekeeper** : woman who looks after a house, does the shopping etc.
4. **sheets** [ʃi:ts], **towels** ['taʊəlz] : bed coverings, made of linen. Pieces of thick material to dry your face and body after washing.

'Now look in *my* bag, Old Joe,' she said, 'and tell me how much you'll give me.'

Old Joe counted up the money for each thing in the box and the bag and wrote some numbers on the wall.

'That's how much I'll give you,' he said. 'And no more. I always give too much and so I'm poor.' Then he opened a dirty bag and put the money on the floor. 'When he was alive, he frightened people and they hated him. So we get the profits now that he's dead. Ha, ha, ha!'

Scrooge watched this in horror. 'Spirit! I see and I understand. This could happen to me. Oh God, what's this now?'

The scene changed and he was near a bed. It had no blankets or curtains. There was only an old sheet with something under it – the body of a dead man. The Ghost pointed at the head, but Scrooge couldn't pull down the sheet and look at the dead man's face. He was shaking [1] with terror. The body was cold, rigid, and alone in that dark room. 'How terrible!' thought Scrooge. 'Not a man, woman or child to say that he was kind to them in life and to remember him with love!' Then he heard the sound of rats behind the walls. Were they waiting, were they going to jump on the bed and...?

'Spirit!' he said. 'What a horrible place! I'll always remember this scene. Can we go now?'

But the Ghost still pointed at the dead man's head.

'I understand,' Scrooge said. 'But I can't do it. I ask you to show me somebody who is sorry that this man is dead.'

The Ghost took him to Bob Cratchit's house. The mother and children were sitting round the fire. They were quiet, very quiet. The little Cratchits sat like statues in a corner. Peter was reading.

---

1. **shaking** ['ʃeɪkɪŋ] : trembling.

'When is Father coming?' he asked. 'He's late. But I think he walks slower now.'

'I remember when he walked very fast with – with Tiny Tim on his shoulder,' said the mother. 'But Tiny Tim was very light – and his father loved him so much. Ah, there's your father at the door!'

Bob came in. He drank some tea while the two little Cratchits put their faces close to his, saying, 'Don't be sad, Father!'

So Bob tried to be cheerful; but suddenly he cried, 'My little child! My little boy!'

He went to a room upstairs. It looked as bright and happy as Christmas. He sat on a chair next to the bed. There was a little child on it. It was Tiny Tim, and he wasn't sleeping. He was dead. Bob kissed the little face; then he went downstairs.

'I met Mr Scrooge's nephew in the street,' he told the family. 'He asked me why I was so sad. When I told him, he said he was very sorry and wanted to help us. I think he's going to find a job for Peter.'

'He's a very good man,' said Mrs Cratchit.

'Yes. Children, when you all leave home in a few years, you won't forget Tiny Tim, will you?'

'Never, Father!' they all cried.

'Thank you. I feel happier now,' Bob said.

Scrooge said to the Ghost, 'Oh, please tell me who that dead man was!'

The Ghost took him near his office, but it didn't stop.

'Wait!' said Scrooge. 'My office is in that house. Let me go and see what I'll be in the future.'

The Ghost continued walking. Scrooge ran to the window of his office and looked in. He saw an office, but it wasn't his. Everything was different, including the man at the desk. He followed the Ghost again. It stopped at the gate of a cemetery.

A Christmas Carol

'Am I going to learn the dead man's name now?' asked Scrooge.

The Spirit led him to a grave. [1] He went near it, trembling.

'Before I look at the name,' he said, 'answer me one question. Is it really necessary for these things to happen or are they only possible?'

The Ghost didn't answer.

'I mean, if men change their lives and become better, will the future change too? Is this what you want to tell me?'

The Ghost was silent. Scrooge went slowly towards the grave, still trembling. He read the name on the gravestone: EBENEZER SCROOGE.

He fell on his knees. [2] '*I* was the dead man in the bed! Oh, Spirit! Oh no, no! Listen, I've changed. I won't be the same man as before. Tell me there is still hope – please! Tell me that if I change my life, the things that you have shown me will be different!'

The Spirit's hand trembled.

'I will celebrate Christmas with all my heart!' Scrooge continued. 'And I'll always try to have the Christmas spirit – every day of the year! I will live in the past, the present and the future. I will not forget the lessons that they teach. Oh, tell me that I can clean the name off [3] this stone!' [4]

Scrooge held up his hands to the Ghost but suddenly it vanished. There was only a bed-curtain in front of his eyes.

1. **grave** : tomb, burial place.
2. **knees** [ni:z] :
3. **clean the name off** : remove the name from.
4. **stone** : (here), tombstone.

Understanding the text

FCE 1 **For questions 1-10 choose the answer (A, B, C or D) which you think best fits according to the text.**

1. The Ghost of Christmas Yet To Come frightened Scrooge because
   A ☐ it was tall.
   B ☐ it was looking at him.
   C ☐ it wanted to help him.
   D ☐ it wore black and it was silent.

2. The three businessmen at the Exchange
   A ☐ were not sad that Scrooge was dead.
   B ☐ wanted to have lunch.
   C ☐ didn't know that Scrooge was dead.
   D ☐ didn't know Scrooge.

3. Scrooge
   A ☐ didn't know the two rich businessmen.
   B ☐ wondered why he was not there in the future.
   C ☐ knew who the two rich businessmen were talking about.
   D ☐ trembled because the Ghost was showing him the future.

4. Old Joe
   A ☐ bought and sold things.
   B ☐ probably gave the women a lot of money.
   C ☐ didn't want to buy anything.
   D ☐ didn't know where the things came from.

5. The two women
   A ☐ were Scrooge's friends.
   B ☐ were selling some of their old things.
   C ☐ worked for Old Joe.
   D ☐ stole the things from Scrooge's house.

6. The Ghost

    A ☐   wanted Scrooge to sleep in the bed.

    B ☐   wanted Scrooge to look at the dead man's face.

    C ☐   wanted Scrooge to see the rats.

    D ☐   wanted Scrooge to feel sorry for the dead man.

7. Bob Cratchit

    A ☐   wasn't sad when he came in.

    B ☐   thought that Tiny Tim was sleeping.

    C ☐   asked his children to remember Tiny Tim.

    D ☐   told Fred that Scrooge was dead.

8. The Ghost took Scrooge

    A ☐   into his office.

    B ☐   to a desk.

    C ☐   to a cemetery.

    D ☐   to his house.

9. At the grave Scrooge asked the Ghost

    A ☐   to help him become better.

    B ☐   its name.

    C ☐   if it was his grave.

    D ☐   if men could change the future.

10. The Ghost said

    A ☐   there was still hope for Scrooge if he changed.

    B ☐   nothing.

    C ☐   goodbye to Scrooge.

    D ☐   that Scrooge must clean the name off the gravestone.

2 **Complete the questions with the correct interrogative word. Then match the questions to their answers.**

a. .... were the businessmen talking about?

b. .... did Old Marley die?

c. .... was Bob Cratchit sad?

d. .... was Old Joe's shop?

e. .... did Scrooge see in the cemetery?

f. .... did Scrooge feel when he saw the black ghost?

1. ☐ Because Tiny Tim was dead.

2. ☐ Very frightened.

3. ☐ His grave.

4. ☐ Scrooge.

5. ☐ Seven years ago.

6. ☐ In a poor part of the city.

3 **Look – the Ghost is speaking to Scrooge and the old man is making a lot of promises! Complete the sentences with *will* or *won't* and the appropriate verb where necessary.**

Ghost: Well, Ebenezer Scrooge? Do you really want to clean the name off the gravestone?

Scrooge: Oh yes! I ⁰ ..*will be*............. a different man – I promise! I ..*won't be*.......... angry with people.

Ghost: How ¹ .... you ² .... your life?

Scrooge: Well, I ³ .... kind and generous to people and I ⁴ .... stay away from them any more. When Fred invites me to his house, I ⁵ .... no again.

Ghost: What about the poor people?

Scrooge: Oh, I ⁶ .... them a lot of money and I ⁷ .... help Bob Cratchit's family. I ⁸ .... like a second father to Tiny Tim! I ⁹ .... the same man as before.

Ghost: And what **10** .... you do about Christmas?

Scrooge: Well, I **11** .... Christmas is a humbug again and I **12** .... always celebrate it.

Ghost: That's good. I hope that you **13** .... the lessons that you have learnt, Mr Scrooge.

Scrooge: Oh, no I **14** .... ! I **15** .... with the Spirits of the past, present and future in me!

 4 **For questions 1-10, read the text below. Use the word given in capitals at the end of each line to form a word that fits in the space in the same line. There is an example at the beginning (0).**

The ghost took Scrooge to a poor part of the city, where the streets were dirty and narrow.

There were dark shops and houses. The people looked ugly and **(0)** .unhappy......... .	HAPPY
A lot of them were **(1)** .... In fact, the quarter was very **(2)** .... There were no policemen, and it was full of **(3)** .... Scrooge followed the ghost into the **(4)** .... shop in the street.	DRINK DANGER CRIME DIRTY
A grey- **(5)** .... old man called Old Joe sat with a woman.	HAIR
They were taking some of Scrooge's **(6)** .... from a box.	POSSESS
The woman, who was Scrooge's **(7)** ...., said that he had been a **(8)** .... old miser, so she had taken some of his things.	CLEAN SELF
When Old Joe gave the women some money, he said that his **(9)** .... had made him poor. But everybody could see that really his business was **(10)**	GENEROUS PROFIT

Prediction

5 A Answer these questions. Give reasons for your ideas.

Do you think...
- Scrooge will really be a different man in future?
- Tiny Tim will die?
- Fred will find a job for Peter?
- Scrooge will die alone and unloved?

B What will Scrooge do to try to be a better man?

....

....

....

....

Writing

6 Some people decide to change and be better in the New Year. They make New Year resolutions, like this:

- I'm going to be good to my teacher.
- I'm not going to watch so much TV.

Write a letter of between 120-180 words in an appropriate style to a friend telling him/her about your resolutions for the New Year.

Dear ....

We had a very good Christmas, and I hope you and your family had a good one too. Well, now that we are in a new year again, I've decided there are some things that I want to change about my life. So I've made some New Year resolutions ....

I've made lots of resolutions before but I couldn't keep them! What about you?

London in Dickens's Time

During the 1840s London quickly became a very big city. In 1850 there were a quarter of a million more people than in 1840. In the area of St Giles 2,850 people lived in only 95 small houses – about 30 people in each house!

In these poor areas, conditions were terrible. The streets were very dirty, and people drank and cooked with bad water from the River Thames. In 1847, 500,000 people died of typhus fever in one month. In winter the fog was often very thick; in summer the smells were horrible.

In one of the most famous areas, Seven Dials, there was a street of second-hand clothes shops called Monmouth Street. Dickens

Seven Dials from *London a Pilgrimage* (1872)
by Gustave Doré and Blanchard Jerrold.

describes it in *Sketches by Boz*, and Old Joe has his shop there in *A Christmas Carol*. The Seven Dials area was full of thieves and poor street-sellers. Many of the old buildings were cleared away when Charing Cross Road and Shaftesbury Avenue were built.

Children aged only nine worked in factories for many hours a day. A lot of very poor people went to workhouses, where they worked very hard in bad conditions, or they lived in very squalid conditions in prison. In 1837, between 30,000 and 40,000 people were in prison for debt.

Victorian London was very busy, crowded, and noisy. It was the capital of a country in the middle of a scientific and economic revolution, and which was creating a huge empire across the world.

The Bank and Royal Exchange, London (1887) by William Longsdail.

1 **Answer these questions.**

1. London's population increased by a quarter of a million
   a. ☐ in 1840.
   b. ☐ in ten years.
   c. ☐ in 1850.

2. In the area of St Giles there were
   a. ☐ too many people.
   b. ☐ about 30 people in 95 houses.
   c. ☐ too many small houses.

3. In 1847 people died of typhus fever because
   a. ☐ the streets were dirty.
   b. ☐ the River Thames was bad.
   c. ☐ they drank and cooked with bad water.

4. Old Joe's shop was
   a. ☐ in Charing Cross Road.
   b. ☐ based on a real street that Dickens knew.
   c. ☐ in *Sketches by Boz*.

5. The poor people's houses in the Seven Dials area
   a. ☐ don't exist today.
   b. ☐ are in Shaftesbury Avenue.
   c. ☐ are now second-hand shops.

6. Children worked
   a. ☐ for nine hours a day.
   b. ☐ in prison.
   c. ☐ at the age of nine.

7. Thirty to forty thousand people were in prison because they
   a. ☐ were dangerous criminals.
   b. ☐ couldn't pay back money to other people.
   c. ☐ couldn't find work.

8. Victorian London was
   a. ☐ in the middle of the country.
   b. ☐ the capital of a growing, prosperous country.
   c. ☐ a quiet place.

Writing

 Last month you read an interesting book about Dickens's London. Write a letter to a pen friend who is going to visit London for the first time. In 120-180 words tell him/her what London was like in Dickens's time.

> Dear ....
>
> I was delighted to hear that you're going to visit London soon. There are lots of beautiful areas to see, but London wasn't always such a pleasant city. I have just read a book ....
>
> Please write and tell me about your visit!

INTERNET PROJECT

Let's find some more information about some of the places described in the dossier London in Dickens's Time.

Organise your class into two groups and each group can report to the class on one of the following:

▶ Seven Dials, Monmouth Street, Shaftesbury Avenue and Charing Cross Road in Dickens's time.

▶ Seven Dials, Monmouth Street, Shaftesbury Avenue and Charing Cross Road today.

What differences did you find? Now go back to the Web and find some more information about how to get there, where to stay and what to do during a weekend spent in this part of London.

Now write your itinerary and describe it to the class.

Before you read

FCE **1** Read the text and think of the word which best fits each space. Use only one word in each space. There is an example at the beginning (0).

The bed was his, the room was his, and best (0) .*of*............... all, he still (1) .... time to be a better man. He jumped (2) .... of bed.

'I will (3) .... with the spirits of the past, present and future in me!' he said, on his knees and with tears in his (4) .... 'Thank you, Jacob Marley! God bless Christmas!' Then he (5) .... on his clothes.

'My clothes are here: I am here. But the future is not here (6) .... and I can change it!' he said, laughing and crying (7) .... the same time. '(8) .... shall I do first? Oh, I feel as (9) .... as a feather! I'm as happy as (10) .... angel! A Merry Christmas to everybody! A Happy New Year to (11) .... the world!'

He danced in the sitting-room and looked (12) ....

'There's the (13) .... where Jacob Marley came in. There's the place where the Ghost of Christmas Present sat. There's the window where I (14) .... the ghosts in the air. It's all right, it's all true, it all happened!'

And he laughed (15) .... laughed. Then the church bells (16) .... - ding, dong, ding, dong! It was a glorious sound! He (17) .... the window and put (18) .... his head. No fog. It was a bright, sunny day, and the (19) .... was cold and sweet.

'What's today?' Scrooge shouted to a boy in (20) .... street.

'Eh?' The boy looked very surprised.

'What's today, my boy?'

'It's Christmas Day.'

Chapter Six

'A Merry Christmas, Mr Scrooge!'

track 07

The bed was his, the room was his, and best of all, he still had time to be a better man. He jumped out of bed.

'I will live with the spirits of the past, present and future in me!' he said, on his knees and with tears in his eyes. 'Thank you, Jacob Marley! God bless Christmas!'

Then he put on his clothes.

'My clothes are here: I am here. But the future is not here yet and I can change it!' he said, laughing and crying at the same time. 'What shall I do first? Oh, I feel as light as a feather! [1] I'm as happy as an angel! A Merry Christmas to everybody! A Happy New Year to all the world!'

He danced in the sitting-room and looked around.

'There's the door where Jacob Marley came in. There's the place where the Ghost of Christmas Present sat. There's the

---

1. **as light as a feather** ['feðə] : very happy. 'Light' is the opposite of 'heavy'; feather :

window where I saw the ghosts in the air. It's all right, it's all true, it all happened!'

And he laughed and laughed. Then the church bells rang – ding, dong, ding, dong! It was a glorious sound! He opened the window and put out his head. No fog. It was a bright, sunny day, and the air was cold and sweet.

 'What's today?' Scrooge shouted to a boy in the street.

'Eh?' The boy looked very surprised.

'What's today, my boy?'

'It's Christmas Day.'

'Christmas Day!' Scrooge said happily to himself. 'So the Spirits did everything in one night. Hey, boy! Do you know the butcher's shop in the next street?'

'Of course!'

'You're an intelligent boy! Do you know if they've sold that big turkey in the window?'

'You mean the one as big as me?'

'What a nice boy!' said Scrooge. 'Yes, that one.'

'No. It's still there.'

'Is it? Oh, good! Go and buy it for me, will you? Tell them to bring it here. If you come back in five minutes, I'll give you half-a-crown.'[1]

The boy ran as fast as possible to the shop.

'I'll send it to Bob Cratchit,' Scrooge said. 'Ha, ha! He won't know who sent it!'

And he wrote Bob's address on a piece of paper. When the butcher's man arrived with the enormous turkey, Scrooge told him to call a cab.[2] He paid for the turkey and the cab, and he gave

---

1. **half-a-crown** : two shillings and sixpence – in Dickens's time, a lot of money!
2. **cab** : (here) taxi, pulled by a horse.

the boy half-a-crown. He was laughing all the time. Then he put on his coat and walked along the street. He looked at all the people with a happy smile.

'Good morning!' people said to him. 'A Merry Christmas to you!'

And Scrooge answered in the same way.

'Ah, there are the two gentlemen who were asking for money in my office yesterday,' he said. 'How do you do, my dear sirs! A Merry Christmas to you!'

'Mr Scrooge?' asked one of them.

'Yes. That's my name and perhaps you don't like me. Please excuse me for yesterday. Listen...'

Scrooge spoke quietly in the man's ear.

'Are you serious, Mr Scrooge?' The man was very surprised.

'Of course. Can you do me that favour?'

'My dear Scrooge, that's very generous! I don't know what to say to thank you.'

'Don't say anything. Come and see me tomorrow. I'll give it to you then. All right?'

Then Scrooge continued walking. He watched the people, he kissed little children, he played with some dogs, he looked at everything with love, and he felt very happy.

In the afternoon he went to his nephew's house. Everybody was in the dining-room. [1]

'Fred!' said Scrooge at the door.

'My God! Who's that?' cried Fred.

'It's your Uncle Scrooge. I have come to dinner. Can I come in, Fred?'

'Come in? Of course, uncle! You're very welcome.'

---

1. **dining-room** ['daɪnɪŋ ruːm] : place where people eat.

Everybody was happy to see Scrooge and he was happy to see them. They ate a wonderful dinner, and then they played wonderful games and had a wonderful time!

Scrooge was in his office early the next morning. He was waiting for Bob Cratchit. Of course, he knew Bob would be late. Nine o'clock and no Bob. A quarter past nine. No Bob. At nine-twenty Bob ran in. He went into his room immediately and started to work fast.

'Hallo!' Scrooge said in his old angry voice. 'You're late!'

'I'm very sorry, sir!' Bob answered.

'Are you? Come here, Cratchit.'

'It's only once a year, sir,' said poor Bob. 'It won't happen again.'

'Well, my friend, I hope not,' Scrooge said with a big smile. 'Because I'm going to give you a bigger salary!'

Bob trembled. He couldn't believe his ears.

'A merry Christmas, Bob!' said Scrooge. 'This will be your happiest Christmas! Yes, I'm going to give you a lot more money and I'm going to help your poor family. Come on, make a very big fire and let's have a drink, Bob Cratchit!'

Scrooge gave Bob more money, helped his family and did much more. Tiny Tim did NOT die and Scrooge was a second father to him. He became a good friend, a good manager and a good man. A few people laughed at him, of course, but he knew that some people always laugh at anything new, strange, and good. *He* often laughed now, and that was the most important thing to him. He didn't see the Spirits again, and he celebrated every Christmas with all his heart. And, like Tiny Tim, he said, 'God bless everyone!'

Understanding the text

1 **What did Scrooge do on Christmas Day? Put the sentences in their correct order.**

a. ☐ He asked a boy in the street to go and buy a turkey.

b. ☐ He felt very happy because he could still change the future.

c. ☐ In the afternoon he went to eat at his nephew's house.

d. ☐ Scrooge jumped out of bed.

e. ☐ Then he went into the street and spoke to two gentlemen.

f. ☐ He danced into the sitting-room and opened the window.

g. ☐ He thanked Jacob Marley and put on his clothes.

h. ☐ Next morning he told Bob that he was going to give him more money.

i. ☐ He sent it to Bob Cratchit.

j. ☐ They were very surprised because he wanted to give lots of money to the poor.

2 **Can you complete these sentences with information from Chapter Six?**

a. On Christmas morning the weather ....

b. The boy in the street was very surprised because ....

c. When Scrooge walked along the street, he ....

d. Scrooge asked the two gentlemen to ....

e. At Fred's house everybody ....

f. When Bob came in at 9.20, he ....

g. Bob couldn't believe his ears when ....

h. Some people laughed at Scrooge because ....

3 You have now read all the chapters. What do you remember about the characters? Of which of the people (A-F) are the following true? There is an example at the beginning (0).

A Scrooge   **D** Bob Cratchit

B Mrs Cratchit   **E** Tiny Tim

C Fred   **F** Fred's wife

Who...

0. [C] invites his uncle to dinner on Christmas Day.

1. ☐ thinks Scrooge is stupid and bad.

2. ☐ sends a turkey to his clerk.

3. ☐ walks with a small crutch.

4. ☐ doesn't want to drink to Scrooge.

5. ☐ thinks of Scrooge for the game of 'twenty questions'.

6. ☐ is late for work on December 26th.

7. ☐ cooks a good Christmas dinner for her family.

8. ☐ doesn't die.

9. ☐ gets the right answer in the game of 'twenty questions'.

10. ☐ changes from a bad old miser to a good generous man.

11. ☐ says he's going to invite Scrooge to dinner every Christmas.

12. ☐ would like to tell Scrooge what she thinks of him.

13. ☐ drinks to Scrooge because it's Christmas.

14. ☐ became a second father to Tiny Tim.

Question tags

'Go and buy it for me, will you?'

Scrooge says 'will you' at the end of the sentence to ask for agreement or confirmation from the boy. ('Yes, I will'). This is a **question tag**. We use question tags when we speak.

After the imperative we use 'will you'.

'Come here, Cratchit, will you?'

'Don't be late, will you?'

After negative statements we use the affirmative interrogative with **do**, **does**, or **did**.

'You don't like Scrooge, do you?'   (present tense)

'Tiny Tim didn't die, did he?'   (past tense)

After positive statements we use the negative interrogative with **don't**, **doesn't**, or **didn't**.

Bob Cratchit loves his son so much, doesn't he?

Scrooge felt very happy, didn't he?

Here are some examples with **to be**, **to have**, and **can**.

The future isn't here yet, is it?

Scrooge can change the future, can't he?

Bob was late for work, wasn't he?

Tiny Tim has got a second father now, hasn't he?

Scrooge is as happy as an angel, isn't he?

The Cratchits haven't got any problems about money now, have they?

4 **Match the sentences in A with their question tags in B.**

A		B	
a. ☐ You can do me a favour,		1.	hasn't he?
b. ☐ Scrooge is a happy man,		2.	have they?
c. ☐ Make a big fire, Bob,		3.	didn't they?
d. ☐ Some people laughed at Scrooge,		4.	wasn't he?
e. ☐ They haven't sold that big turkey yet,		5.	will you?
f. ☐ Scrooge didn't see the Spirits again,		6.	can't you?
g. ☐ The boy has got half-a-crown,		7.	did he?
h. ☐ Fred was happy to see Scrooge,		8.	isn't he?

5 Write question tags for the following.

a. Scrooge was in his office early, ... ?

b. Come and see me tomorrow, ... ?

c. The Cratchits have got enough money now, ... ?

d. They ate a wonderful dinner, ... ?

e. The turkey is still in the window, ... ?

f. Bob didn't know who sent the turkey, ... ?

6 A Let's describe the four ghosts in the story! Complete the 'looked' and 'was' columns with suitable adjectives. Then complete the 'had' column, like the example.

| silent | happy | small | frightening |
| young | tall | old | large |

	looked	was	had
Marley's Ghost	sad	old handkerchief,	boots, chain, round face, cold eyes
First Spirit	young		
Second Spirit			
Third Spirit			

B Now use the information in A to write a summary.

Listening

7 Listen to this extract from page 97 and, without looking at the text, complete the boy's part of the conversation.

Scrooge:	What's today?
Boy:	1 ....
Scrooge:	What's today, my boy?
Boy:	2 ....
Scrooge:	So the Spirits did everything in one night. Hey, boy! Do you know the butcher's shop in the next street?
Boy:	3 ....
Scrooge:	Do you know if they've sold that big turkey in the window?
Boy:	4 ....
Scrooge:	What a nice boy! Yes, that one.
Boy:	5 ....
Scrooge:	Is it? Oh, good! Go and buy it for me, will you?

Now listen to the next part on page 98 and, without looking at the text, complete Scrooge's part of the conversation.

Scrooge:	6 ....
Gentleman:	Mr Scrooge?
Scrooge:	7 ....
Gentleman:	Are you serious, Mr Scrooge?
Scrooge:	8 ....
Gentleman:	My dear Scrooge, that's very generous! I don't know what to say to thank you.
Scrooge:	9 ....

Now, check your answers.

Summary

FCE 1 A **For questions 1-30, read this summary of *A Christmas Carol* and think of the word which best fits each space. There is an example at the beginning (0).**

A Christmas Carol is the story of Ebenezer Scrooge, an
(**0**) .old............ miser who didn't (**1**) .... Christmas. He
was always unkind to people, and he (**2**) .... away from
them.

One Christmas Eve the ghost of Scrooge's dead (**3**) ....
Jacob Marley came to his rooms. Marley wore a long chain
(**4**) .... his body and he (**5**) .... rest because
during his life he only made (**6**) .... He told Scrooge
that three Spirits would (**7**) .... him.

The (**8**) .... Spirit was the Ghost of Christmas Past. It
showed Scrooge (**9**) .... when he was a lonely, unhappy
schoolboy. Then he (**10**) .... himself in the office where
he first started (**11**) .... His manager, Mr Fezziwig, was
kind and (**12**) ...., and he gave a big Christmas party.
Next, Scrooge saw his fiancée Belle, (**13**) .... said that he
loved money more (**14**) .... her. Finally, the Spirit
showed him Belle some years (**15**) .... when she was the
happy mother of a big family. Scrooge felt very (**16**) ....
because he (**17**) .... have a family.

The Second Spirit was the Ghost of Christmas Present. It
(**18**) .... him to his clerk Bob Cratchit's house. Bob was
(**19**) .... so his family had only a (**20**) ....
Christmas dinner, but they all (**21**) .... to Scrooge. When
the Ghost told Scrooge that Tiny Tim would probably
(**22**) ...., he was very unhappy. Then he saw his
(**23**) .... Fred's family. They were all enjoying
themselves, and Scrooge began to (**24**) .... himself too!
(**25**) .... this, the Ghost showed him two thin, hungry
(**26**) ...., who (**27**) .... trembling with cold.
Scrooge was (**28**) .... And he also felt very bad when
the Ghost repeated his own (**29**) .... about workhouses
and (**30**) ....

B **Now finish the summary in your own words. Use these ideas as a guide.**

- the three businessmen at the Exchange
- Old Joe / the first woman / the housekeeper
- the dead body in the bed
- Bob Cratchit's house
- Scrooge's office
- the cemetery
- Christmas Day: Scrooge's feelings / the turkey / the two gentlemen / Fred's house
- December 26th: Bob Cratchit / Scrooge and Tiny Tim / Scrooge at the end of the story

 You are going to read an extract from a book about Charles Dickens and *A Christmas Carol*. For questions 1-6 choose the answer (A, B, C or D) which you think fits best according to the text.

Charles Dickens and *A Christmas Carol*

The idea for *A Christmas Carol* came to Dickens when he visited some schools for poor children in Manchester, a city in the north of England.

Then he created the story while he was walking around 'the black streets of London, fifteen and twenty miles a night' – when everybody was in bed!

Dickens's characters were very real to him. One of his friends remembered that one day when he was walking with Dickens, the novelist said, 'Mr Micawber [1] is coming. Let's go into another street and get away [2] from him.' Dickens wrote that when he was

---

1. **Mr Micawber** : character from Dickens's novel *David Copperfield*.
2. **get away** : escape from (phrasal verb).

writing *A Christmas Carol*, he laughed and cried with the characters. Tiny Tim and Bob Cratchit were always pulling his coat because they wanted him to sit down at his desk and continue writing the story of their lives.

After Dickens died, one of his children called Charley said that the children of his father's brain were '…more real to him than we were.'

The titles of many of Dickens's novels are names (*Oliver Twist, David Copperfield, Martin Chuzzlewit*). These names were very important to Dickens because he knew that they gave life to his characters. The name Scrooge is now part of the English language. If you say, 'That man is a real Scrooge', you mean that he never spends any money.

1. Dickens got the idea for *A Christmas Carol*

    A ☐ in London.

    B ☐ in Manchester.

    C ☐ in bed.

    D ☐ in the street.

2. Dickens walked

    A ☐ thirty-five miles around London.

    B ☐ during the day.

    C ☐ and wrote at the same time.

    D ☐ and created *A Christmas Carol* in his head.

3. Dickens's characters

    A ☐ were more real to him than people.

    B ☐ were like real children.

    C ☐ made him want to run away.

    D ☐ pulled his clothes in the street.

4. Charley was

    A ☐ in one of Dickens's novels.

    B ☐ Dickens's son.

    C ☐ not as real to Dickens as his other children.

    D ☐ a child of Dickens's brain.

5. Oliver Twist is

A ☐ a real person.

B ☐ a character in *David Copperfield*.

C ☐ the title of a Dickens novel.

D ☐ not an important name to Dickens.

6. The name Scrooge

A ☐ gives life to the English language.

B ☐ hasn't got any meaning.

C ☐ is the name of a real person.

D ☐ means somebody who is mean with money.

3 **Read the text and answer the questions.**

When I think of Christmas, I always think of Dickens. Christmas without Dickens wouldn't be Christmas for me. Snow in the street, bright fires and decorations and Christmas trees in the house; lots of good food and drink and presents; carols and games and pantomime. Everybody is very kind and happy – you know, a real 'Dickensian' Christmas!

But I know it's only a dream. Christmas was very different then and it's different now! I think Dickens invented a perfect Christmas. He was saying: 'Look! Christmas can be a very happy time if you want!' Dickens loved Christmas; to him it was a very special time – a time, above all, for children. He invented lots of games, he organised music and dancing and singing, and he did conjuring tricks.

But for most people Christmas Day was a day of rest because there was only one day's holiday. Of course, they had Christmas lunch, some games, and music perhaps. But Christmas wasn't commercialised like our modern Christmas. In Dickens's time it wasn't big business. People didn't go shopping weeks before, they didn't spend so much money – and they didn't watch TV. Yes, today we don't play Christmas games or sing at home. We watch TV.

Write some ideas for the categories below.

a. 'Dickensian' Christmas: snow, bright fires, decorations, trees ......

....

....

b. Dickens's Christmas: a special time for children ....

....

....

c. Christmas in Dickens's time: a day of rest ....

....

....

d. Christmas today: ....

....

....

T: GRADE 7

4 **Try to give a brief account of the content of *A Christmas Carol.* Now choose a chapter that you particularly enjoyed and explain it in more detail.**

T: GRADE 7

5 **Topic – Money**

In *A Christmas Carol* we see Scrooge as a mean old miser at the beginning of the story and then as a kind, generous man at the end. Describe how Scrooge feels at the end of the story. What, in your own words, is the moral of this story? In Britain we tell similar stories or fables to our children. Do you have stories or fables like this in your country? Try to find an example and tell the class about it.

Now think about Scrooge's attitude to money. Do you save money like Scrooge or do you spend it as soon as you get it?

Exit Test Key

1A Summary

1. like 2. kept, stayed
3. partner, friend 4. around
5. couldn't 6. money 7. visit
8. First 9. about, himself
10. saw 11. working
12. generous 13. who 14. than
15. later, after 16. sad
17. didn't 18. took 19. poor
20. small 21. drank 22. die
23. nephew 24. enjoy 25. After
26. children 27. were 28. sad,
upset 29. sentence, comment,
phrase 30. prisons

1B Open answers. Possible : The Third Spirit was the Ghost of Christmas Future. It wore long clothes and a black hood so that Scrooge couldn't see its face. It showed three businessmen in the City of London discussing someone's death. Scrooge wondered who they were talking about, they didn't seem to like this person very much. It then took Scrooge to a very poor part of the city and to a shop owned by a man called Old Joe. A woman came into the shop and sold Old Joe the contents of her bag. Then a housekeeper did the same. They both said that the things belonged to an old miser who had died.
Then the Spirit took Scrooge to Bob Cratchit's house. Tiny Tim was dead and everyone was very sad. They all promised never to forget Tiny Tim. The Spirit and Scrooge walked straight past Scrooge's office and then to the cemetery. Scrooge was confused, he wanted to see himself in the future but the Spirit said no and Scrooge read his own name on a gravestone. Here lies Ebenezer Scrooge. No-one cared or missed Scrooge. He wanted to change his ways.
The next day Scrooge woke up happy. He called to a boy in the street. 'Go and get me that turkey from the shop in the next street'. He sent the turkey to Bob and his family and then promised the men some money for the poor. Scrooge ate with his nephew Fred and had the best Christmas ever. On Boxing Day he promised Bob a bigger salary and then helped his family and Tiny Tim who did not die. From that day Scrooge always celebrated Christmas.

2 1. B 2. D 3. A 4. B 5. C
6. D

3 a. lots of good food and presents, carols and games and pantomime.
b. a perfect Christmas, he invented games, organised music and dancing and singing.
c. only one day's holiday, Christmas lunch, some games.
d. commercialised, everyone spends a lot of money, no games, TV.

4 Trinity
Open answer.
5 Trinity
Open answer.